THE TRIVIA LOVER'S

GUIDE TO THE

WORLD

GUIDE TO THE WORLD

THE TRIVIA LOVER'S

LOVER'S

GEOGRAPHY *for the* LOST *and* FOUND

GARY FULLER

FALL RIVER PRESS

New York

FALL RIVER PRESS

New York

An Imprint of Sterling Publishing
387 Park Avenue South
New York, NY 10016

© 2012 by Rowman & Littlefield Publishers, Inc.
All maps by Matthew Renault

This 2014 edition printed by Fall River Press by arrangement with
Rowman & Littlefield Publishers, Inc.

ISBN 978-1-4351-4709-6

Distributed in Canada by Sterling Publishing
c/o Canadian Manda Group, 165 Dufferin Street
Toronto, Ontario, Canada M6K 3H6
Distributed in the United Kingdom by GMC Distribution Services
Castle Place, 166 High Street, Lewes, East Sussex, England BN7 1XU
Distributed in Australia by Capricorn Link (Australia) Pty. Ltd.
P.O. Box 704, Windsor, NSW 2756, Australia

For information about custom editions, special sales, and premium
and corporate purchases, please contact Sterling Special Sales
at 800-805-5489 or specialsales@sterlingpublishing.com.

Manufactured in the United States of America

2 4 6 8 10 9 7 5 3 1

www.sterlingpublishing.com

This book is dedicated to Barbara Fuller, whose creative and editorial assistance is gratefully acknowledged.

CONTENTS

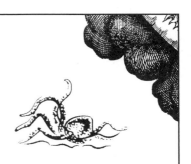

MAPS

ACKNOWLEDGMENTS

Thanks go to the following for reading and commenting on early drafts: Michael Fuller, Teresa Reddekopp, Kathleen Simi, John Fuller, and Matthew McGranaghan.

Inspiration for the book was provided by Ian Reddekopp, Alexandre Fuller, Emma Reddekopp, Hannah Simi, Isaac Reddekopp, Simon Fuller, Elliot Reddekopp, Sophie Bartels, Aidan Reddekopp, Sebastien Fuller, and Caleb Simi.

I am grateful to Jim Metzdorf for technical assistance.

INTRODUCTION

By definition, trivia is unimportant! Yet despite its seeming un-importance, trivia games of one sort or another are very popular. They're played at family gatherings, as fund-raisers for charitable groups, as big-prize television shows, among passengers on cruise ships, sometimes even at school. Geographic trivia questions seem to be especially popular and, often, especially difficult. Just as geography itself has always appealed to people who wonder what's over the next hill, geographic trivia reminds us of places we've visited or hope to visit someday. So, we challenge ourselves with difficult questions because it's entertaining and enlightening to do so.

Trivia, if given the right context—the right setting—can be magi-cally transformed into something both important and interesting. In this book, I ask and answer 150 trivia questions, but I also provide a setting and an explanation for each answer. If you're not very good at trivia, you'll certainly get better by reading this book. Not only will you be able to answer 150 reasonably difficult questions, you'll find the explanations will allow you to answer many more—and make up a few of your own! If you're already a master of trivia, the explana-tions, maps, and tables will not only add to your mastery but expose you to ideas that you haven't thought about before.

Trivia games sometimes turn into family room warfare, particu-larly when an argument develops over the correct answer. I'm pretty sure all the answers I've given were correct when I wrote them, but facts sometimes do change. If you have a complaint about an answer, email or write me in care of the publisher. Also, if you have a particularly good geographic trivia question, send it to me, and I may be able to include it in a future edition.

THE EFFECTS OF
GEOGRAPHIC IGNORANCE
ON THE MODERN WORLD

Pat Sajak of television's *Wheel of Fortune* seems about as unflappable as a game show host can get. With thousands of shows under his belt, what could possibly happen that was new? The category on the show one evening was "Where are you?" The puzzle consisted of three words that described a place, and there was a $1,000 bonus for naming the place once the puzzle was solved. A contestant solved the puzzle with minimal difficulty. The three words were: *pineapple, aloha, lei*. After some hesitation, the contestant offered "Idaho" as the answer. There was applause in the audience from those who apparently thought pineapples were a principal crop just outside Boise, but Pat Sajak was momentarily speechless! By the way, if you happen to be lost geographically, the correct answer was "Hawai'i."

Every television quiz show in the United States reveals the same thing: geographic ignorance is widespread. On *Jeopardy!* and *Who Wants to Be a Millionaire*, viewers can watch brilliant people answer obscure questions but then wash out when they confuse Switzerland and Sweden or think the Nile River is in Asia. This situation wouldn't be so bad except geography is pretty important stuff. It helps us understand how the world is organized—how history, politics, economics, climate,

1

and culture coalesce in specific places. Just as music has a very basic vocabulary of *notes*, so geography's vocabulary is made up of *places*. Without those places, it's hard to understand religion and language and much of anything else that makes our world what it is. Most of the literature we enjoy—from the classics to modern pulp fiction—is enriched by a basic understanding of the places where the stories are set.

If geographic ignorance were limited to game shows, there would be no need for this book. Unfortunately, it's sometimes where you'd least expect it. A few years ago I was privileged to attend a meeting of State Department personnel. People chosen for State Department positions are highly educated, often from the best colleges and universities, but sometimes lack basic geographic information and understanding. During a lull in the proceedings, an informal discussion began about unusual languages. A number of interesting languages were mentioned, as one would expect from the collective knowledge at the table. Eventually they turned to me, as an outsider and the only geographer. I nominated Bislama, the pidgin language of Vanuatu. I told them the language developed as a result of the sea slug trade between the French and the Chinese and that the name of the language actually came from the French *beche de mer,* the sea slug itself. Hours later, after the meeting concluded, several participants congratulated me on my imaginative language story. Apparently nobody believed me; they assumed even the country Vanuatu had been made up. I couldn't convince them otherwise.

So what? Who cares about Vanuatu? Despite its collective talent, the US government has an alarming history of knowing next to nothing about areas it considers unimportant and then, *oops*, they become really important. Places such as Vietnam, Somalia, Rwanda, and Afghanistan seem like good examples, but in the past places such as Guadalcanal, Iwo Jima, Tarawa, and, yes, the New Hebrides, or Vanuatu, also come to mind. Ever since the world first shrunk dramatically in 1869 with the opening of the Suez Canal, we've seen places grow ever closer and ever more important to each other . . . and to us. So, yes, Vanuatu is important both to us and to the people that make it their home, even if the folks who work at **FOGGY BOTTOM** are unaware of it.

> Dr. G says: Foggy Bottom is an old neighborhood in Washington, DC, alongside the Potomac River. It's the home of the Harry S. Truman State Department Building. *Foggy Bottom* is a term often used to indicate the State Department itself, particularly by its critics!

The most alarming place for geographic ignorance, however, is the college classroom. Geography is rarely taught in our primary and secondary schools. Sure, there are localized exceptions, the occasional school or teacher trying to offer a little geography. Without an organized curriculum and an understanding of major concepts, however, teachers usually offer little more than an exercise in pointless memorization. The average freshman college student is among the most geographically lost of all. When I was teaching introductory geography at Penn State, Ohio State, and the University of Hawai'i, I would occasionally ask students to locate the country of New Zealand on an outline map of the world, one that showed national boundaries but not the names of countries or other places. There wasn't a corner of the world that failed to be labeled "New Zealand." **Answer 1:** Fewer than 15 percent of the students could find it (in the southwestern Pacific Ocean east of Australia), and a small number even located it in the space reserved for the United States. It's fortunate that we're reasonably good friends with New Zealand, because I doubt we could find it if we decided to take military action against it.

Knowing the location of New Zealand is simply a vocabulary matter, like knowing what *bailiwick* means or where middle C is on the piano keyboard. It's not high-order knowledge, but you have to start somewhere!

I think colleges and universities could help the situation a bit by not describing their locations to the students they accept. Imagine

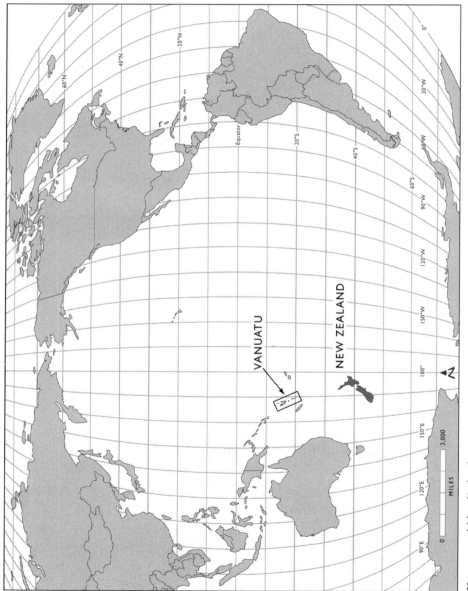

Vanuatu and New Zealand

how panicked students would be if they were ready to leave for college and didn't know where to go. Where is Reed or Trinity, Arizona or TCU? Is there a city called William and Mary or Howard? Worst of all, is **PENN STATE** in Philadelphia or Pittsburgh, or maybe Harrisburg? What fun to have college admissions based on geographical Darwinism!

Dr. G says: Pennsylvania State University is located in University Park, Pennsylvania, but that may be hard to find on a map! University Park is located right next to State College, in almost the exact geographic center of Pennsylvania.

WE THINK BIG, SOMETIMES TOO BIG

Question 2a: What is the first foreign country you would encounter if you went due south from downtown Detroit, Michigan?

Question 2b: Which city is farthest west?: Chicago, Illinois; Denver, Colorado; Reno, Nevada; Los Angeles, California

We don't always agree on how to acquire knowledge, but we generally agree it's important to acquire it. How else can we survive? Our ancestors wondered about what was safe to eat, how best to hunt, how to keep warm in the winter, and millions of other things that enabled them to survive. Today's definition of *survival* by my college students includes employment, and, for that, too, knowledge is necessary.

Viewing the whole gamut of human history, we see that we've acquired knowledge by three means: (1) from divine revelation (what the gods have told us); (2) from authority figures (first our parents and, later, those who seem to know more than we do and are willing to pass their knowledge on to us); and (3) from careful, systematic observation of the world around us, or from what loosely can be called the scientific method. At any point in history, these systems are in fierce contention with each other. Currently, for example, issues associated with evolution pit scientific findings against religious scripture (in the view of some).

Over the past few centuries, we have increasingly come to rely on the scientific method simply because it has proved to be more reliable than the other means of acquiring knowledge. Religion, in most parts of the world, no longer offers explanations for weather phenomena like thunder and lightning and, while we respect noted authorities and experts, that doesn't always work out. The collapse of financial institutions like Lehman Brothers (in the wake of the subprime mortgage crisis) tells us that relying on "informed" opinion is not always a good idea.

Knowledge derived from the scientific method depends ultimately on facts—basic units of what we can see or otherwise detect around us. Geographic facts are collected and added to our memories in the form of "mental maps," which grow in detail and accuracy as we acquire more information. As a general rule, for example, the longer that people live in an area, the more local street names they know.

There are at least two big problems with the collection of facts for our mental maps. One is that the facts change so fast they may be outdated as soon as we form our mental maps. Even small cities like Las Vegas, for example, seem to undergo such rapid change that finding your way around is confounding, even to frequent visitors. Some older people who actually studied geography in school are now totally flummoxed by the names of modern counties in, for example, Africa or in the former Soviet Union.

The bigger problem, however, is that there are far too many facts to fit in our mental maps, particularly when the area covered by the maps extends beyond our immediate neighborhood. By the time we reach a map of the world, we're potentially dealing with billions of facts . . . or more. Being drowned in facts is not useful, as it doesn't provide knowledge that helps us survive.

The problem of having too many facts to deal with is handled easily by our minds: we create big clumps of facts that we can attach to our mental maps. Some of these clumps we call "generalizations," which, unfortunately, can mislead us. One generalization in our mental map is that Canada is north of the forty-eight United States, which leads us, erroneously, to the conclusion that every point in Canada is north of every point in the forty-eight United States.

Answer 2a: Many people, when confronted with the proposition that Canada is due south of Detroit, don't believe it, even when shown evidence in the form of a map. About 5 percent of non-Michigan residents who were students in my classes answered this correctly.

Answer 2b: The truth that Reno, Nevada, is west of Los Angeles, California, is similarly disturbing to many. The same "clump" principle holds here: California is west of Nevada, but not every point in California is west of every point in Nevada.

Maps may be as doomed as libraries. Books of all kinds can now be published, stored, and disseminated electronically. Although our attachment to the printed medium will see us through another generation or two, the handwriting is clearly on the wall. Probably the principal use of maps is to find one's way from one place to another, and the use of GPS systems and electronic renditions of maps is all around us. It remains to be seen whether this **NEW SPACE** will make us more or less aware of where we really are and what surrounds us or whether our connections with the outside world will involve only cell phones, email, and social networking.

> Dr. G says: New Space exists only electronically. It's a virtual space that contains connections but no distance. Aunt Shelia in Australia is as close as the Addams family next door. Warfare has already been waged in New Space, and its scope and influence are bound to increase in the future.

A further point about mental maps: we draw them not only with facts but also with our attitudes toward places. A number of people

Detroit and Canada

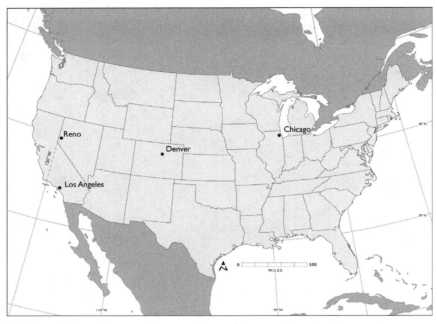

The United States Featuring Chicago, Denver, Reno, and Los Angeles

who are asked the Detroit question will respond in part with a value judgment. One memorable response was, "If I wanted to go to Canada I certainly wouldn't go to Detroit first!" Presumably this reflects the perception that Detroit is an economically depressed, high-crime area.

MAPS ARE FLAT,
THE WORLD ISN'T

Question 3: Excluding Canada and Mexico, what foreign country is closest to the United States?

Even in the absence of a geography curriculum in the schools, most people seem to know that the need to project a spherical world on to a flat surface to produce maps means we get a distorted view of the real world. Despite this understanding, the implications of map projections, for the use of maps generally, are widely overlooked. The result is that maps are wonderful devices to use to tell lies!

The most common world maps in general use for many, many years were Mercator maps or similar projections that were very useful for navigation purposes. Since these projections didn't distort direction, they could be used to travel easily from one point to another by sailing the compass direction from point of origin to point of destination. This wasn't the most direct route to take, but for some journeys it was the easiest to follow.

On Mercator maps, though, the North and South Poles aren't the points they really are but rather lines that are as long as the entire circumference of the world as shown on the map. The closer you go to the poles, then, the more area that has to be shown and the more it's exaggerated. On one Mercator projection that hung in my office, the Antarctic continent is shown to be larger than South America and Africa combined, which it really isn't. Lies told with these maps are pretty obvious. For example, a map showing polar bear habitat in a nature magazine used a Mercator-type projection that greatly exaggerated the extent of the habitat.

More subtle lies are told by what's *not* on maps. If there's a map on the wall in any American elementary school classroom, chances are it's a map based on broad physical features of the world: the oceans are blue, the lowlands are green, the mountains are red, brown, or maybe white if they're really tall. If, however, we're interested in teaching kids about the shrinking world in which we live, doesn't it make sense first to show them where the people are? That's where we find economies, politics, religions, languages, and the host of other things that we associate with humanity. Of course physical features of the earth are important, but it's a question of where we look first.

Many maps of the United States are not accurate (or at least not complete) because they don't include Alaska and Hawai'i. Even professional geographers leave out these states; it's just too much trouble to include these "outliers." The psychological effect of this is interesting and probably important. Both states have serious separatist movements: attempts to leave the Union and opt for some other type of political organization. Tourists to these states, particularly Hawai'i, occasionally inquire whether US currency is accepted. Perhaps of greatest concern is that natives of Hawai'i and Alaska report that they're not accepted as citizens when they move and seek employment in the other forty-eight states. Could better maps change this situation? It wouldn't hurt to try.

Whenever Alaska and Hawai'i are included in maps of the United States, they're stuck in a corner, almost as an afterthought. Alaska is shown at a different scale from the rest of the country and in a way that makes it seem an isolated, small peninsula. **Answer 3**: The result, of course, is that Alaska isn't a part of many peoples' mental maps of the United States, and our close neighbor, Russia, is overlooked. Big Diomede Island in the Bering Strait is the easternmost point of Russia and it's less than two miles from Alaska.

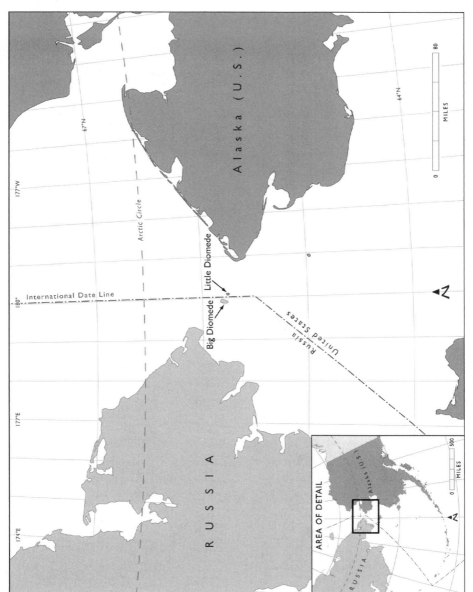

Big and Little Diomede Islands in the Bering Strait

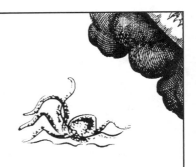

AFRICA BASHING? WHAT'S THAT?

....

Question 4: What is the capital city of Burkina?

....

As an undergraduate student, I had to do a thesis in my senior year to fulfill graduation requirements and earn certification as a secondary school teacher. This was in 1964, and my topic was the independence of Nigeria, an event that had occurred only four years earlier. Every scholarly study and every newspaper and magazine article that I could find at the time were overwhelmingly optimistic about the independence of all of Africa south of the **SAHARA**. Nigeria, with its abundant petroleum resources, was expected to rival Europe's emerging Common Market in terms of overall economic output. Uganda was labeled the "Switzerland of Africa." Sadly, things haven't worked out as expected. The region has been beset by civil wars, military coups, and a general climate of political instability.

> Dr. G says: People who say "Sahara Desert" are being repetitive. *Sahara* is an Arabic word that means "desert"!

It's not popular in the scholarly world or elsewhere to point out the troubles that have befallen Africa. Even public heath reports

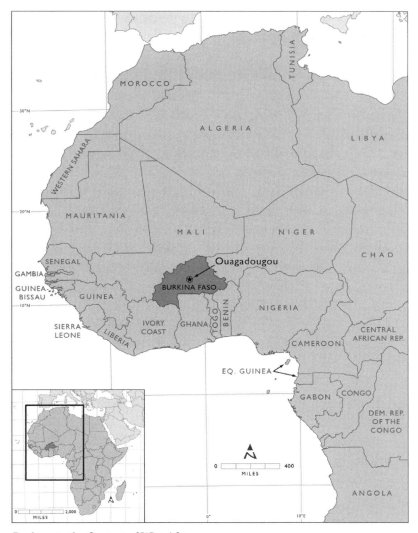

Burkina in the Context of West Africa

on the region, normally among the most objective and politically neutral data that exist, are routinely referred to as "Africa bashing." We could easily dismiss such extreme political correctness as childish foolishness were it not for the fact that ignoring or understating Africa's plight makes things worse. Former president Bill Clinton has stated that he didn't fully appreciate the seriousness of the situation in Rwanda, where possibly 800,000 people died in ethnic violence during his administration. We can share President Clinton's pain when we realize that reports he received were certainly understated by those who feared being called "Africa bashers."

One manifestation of Africa's problems is the numerous name changes of countries. What, for example, happened to Zaire and Biafra? Or, is there really a country named "Burkina"? Indeed there is, but the country was formerly known as Upper Volta and is also officially called Burkina Faso. Our mental maps of Africa have needed frequent revision over the past fifty years!

One fundamental problem that Africa faces is the geography of its national boundary lines. Modern African countries, for the most part, are relics of a colonial past. European powers established colonies in Africa for their own convenience. Boundaries were drawn to accommodate the colonial powers, particularly their transportation needs. Under the thinking at the time, colonies were supposed to provide raw materials to their mother countries and serve as markets for the goods manufactured in Europe. This implied the need for ports and transportation routes, originally river systems, to the interior.

An African colony, thus created to serve the needs of the mother country, might contain several tribes that had been enemies from time immemorial. A given tribe, moreover, might find its traditional lands divided among two or more colonies. Nigeria, the most populous country in Africa, has far more language families than all of Europe, clearly illustrating its tribal diversity. Some African countries, again like Nigeria, have a large Islamic region, a large Christian area, and a relatively lightly populated area that practices traditional religions. *Diversity* is a word that's become a political slogan in the United States, but the diversity that exists in much of Africa is, at present, hostile and disruptive.

Answer 4: Fewer than 1 percent of either my university students or cruise ship passengers know that Ouagadougou is the capital of Burkina, but now you do, too!

PLACES AREN'T ALWAYS WHERE YOU THINK THEY SHOULD BE

Question 5: What large Brazilian city is due south of Chicago, Illinois?

If you imagine going due south from Chicago, you might think that you'd eventually end up somewhere in South America, probably in Brazil given the large size of that country. **Answer 5:** In fact, going due south from many parts of North America, including Chicago, does not get you to Brazil! You would end up in the Pacific Ocean! This question and its correct answer invariably produce widespread disbelief!

The difficulty we have with this question arises because we decided to name a continent "South America" and, while it *is* south of North America, it's also east of most of it. That means that a city on the west coast of South America, like Valparaiso, Chile, is roughly due south of Miami, Florida, on the east coast of North America.

If you had difficulty with this question, you're in good company. In 1493, when Spain and Portugal were battling it out to determine who would claim the new lands uncovered by exploration voyages, **POPE ALEXANDER VI** decided to divide the world between the two seafaring powers. Portugal had colonies on both coasts of Africa, as well as in modern India, Indonesia, and China, and had some interest in Japan as well. The Spanish had new claims in the Caribbean with the possibility of more lands to the west yet to be

17

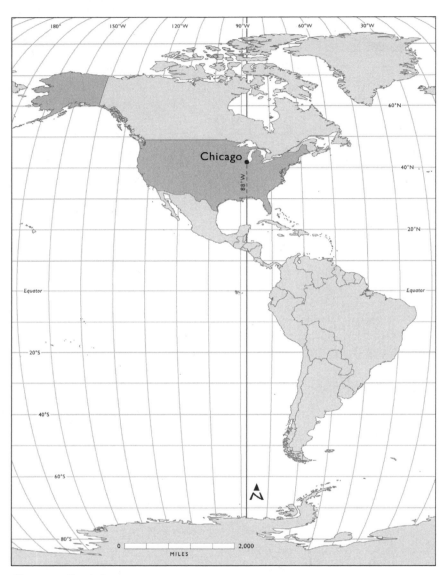

Chicago to South America

discovered, so the Pope decided he'd draw a line running north and south, in the Atlantic Ocean, with Portugal getting everything to the east of the line and the Spanish getting the lands to the west. The line was to run 100 leagues (about 270 miles) west of the Cape Verde Islands.

> Dr. G says: In this case, the Pope issued a papal bull, or edict. The word *bull* comes from the Latin word *bulla*, which is the seal (wax or lead) used to authenticate and close the document.

The Portuguese objected to the line, which to them seemed to give the Spanish way too much in terms of potential claims. The following year, Portugal and Spain agreed to the Treaty of Tordesillas, which moved the demarcation line about 450 miles farther west. Remember, this was only two years after Columbus's first voyage, and the New World had not been explored, much less mapped. The Spanish, as far as they knew, were only granting the Portuguese a portion of the Atlantic Ocean. In fact, they gave Brazil to Portugal, since the whole of South America was much farther east than anyone thought . . . probably. In retrospect, it almost seems that Portugal knew about South America, although there's no record that any of their early voyages along the coast of West Africa strayed far enough west to discover the presence of the South American continent.

If you look at a map showing both Africa and South America, it takes little imagination to see that the two continents were once joined, with Brazil fitting like a puzzle piece into Africa's Gulf of Guinea. An interesting consequence of this is that hot stuff oozed from the fracture between the continents when they separated, so along the coast of Brazil are found what geologists and physical geographers call "intrusive domes." They provide fantastic scenery and help give Rio de Janeiro one of the most beautiful natural settings in the world. Rio's landmark, Sugarloaf Mountain, is an example of these granite domes.

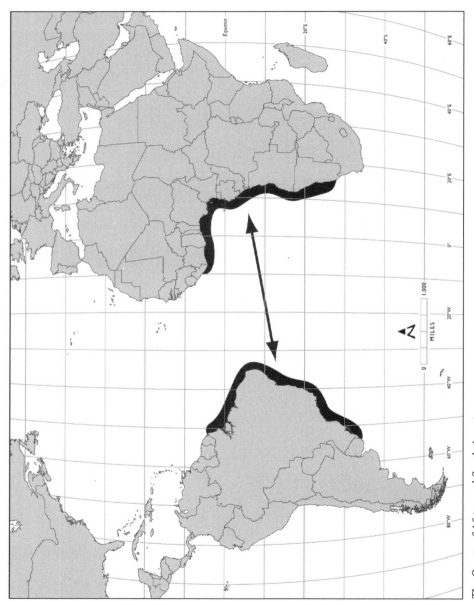

The Coasts of Africa and South America

So the question about what's south of Chicago leads to an explanation of why Portuguese and not Spanish is spoken in Brazil and even why Rio is so attractive!

Being confused about relative location (what geographers call "situation") isn't limited to perception problems with South America. Americans who visit Europe are often unaware of how far north they are: delightful Paris is at roughly the same latitude as much colder Montreal. European settlers in North America had the reverse problem: Plymouth, Massachusetts, is at a latitude roughly the same as Spain's but has a colder climate (as the Pilgrims found out).

CHAPTER 6

STATE CAPITALS AND AMERICAN POLITICS

Question 6a: What is the largest (in population) US state capital?

Question 6b: Is the capital of Kentucky pronounced "Lewis-ville" or "Looey-ville"?

One thing that strikes some parents and teachers as "geographical" is memorizing all the US state capitals. It can be a fun exercise, but it's also difficult and confusing. The capital of Missouri (Jefferson City) seems to be the one many people have a lot of difficulty remembering. Then real confusion arises because New Hampshire's capital (Concord) sounds like it ought to be in Massachusetts, somewhere near Lexington, and because Charleston (the one that's a capital) is *not* in South Carolina but in West Virginia (which also has a Charles Town). Recently, a candidate for the US presidency who was visiting New Hampshire confused the Concord *there* with the one in Massachusetts where the shootin' started the American Revolution (and was heard "'round the world"). In reading the press accounts of this incident, I had the distinct impression that neither she nor most of the American public understood the faux pas!

Knowing all the capital cities in the United States, past and present, would strain even the most phenomenal of memories. Most states or territories have had several capital cities. Boston has the distinction of having been a state or colonial capital for the longest continual period, while Santa Fe is the oldest (and highest) US capital. Some remember that Montgomery, Alabama, and Richmond, Virginia, were capitals of the Confederate States of America but few recall that Danville, Virginia, was briefly the capital after Richmond fell to Grant's army. Even fewer remember that there was a State of Franklin briefly after the American Revolution. Its capital was Jonesboro and it fell by only two congressional votes short of statehood. It's now part of Tennessee.

Throughout the world, capital cities are usually the largest cities in a country. Sure, this isn't true of the United States, Canada, Australia, or Brazil, but by and large, big cities are capitals. (It's not true of New Zealand, either, but since we can't find New Zealand, it probably doesn't matter.) Strangely, however, the state capitals in the United States often are not the largest cities in the state. Montpelier (Vermont) and Carson City (Nevada), for example, are small towns.

Looking at the situation the other way around is more revealing. When an audience is asked to name the largest US state capital, it's common to hear answers like "New York," "Chicago," "Philadelphia," "Detroit," "Los Angeles," "Houston," "Seattle," and "Baltimore." People expect big cities to be the capitals, and none of these large cities are. When these first choices are declared wrong, an audience will begin to advance cities like Boston, Atlanta, and Sacramento, all of which are capitals and reasonably big cities . . . but not the biggest.

We can find a different set of immediate reasons for locating each different state capital, but the distribution of America's population is an important overall factor. The United States was a predominantly rural country when we gained our independence. About 90 percent of Americans lived in rural areas in 1790 and, at that time, the definition of an urban place was a collection of only fifty people! In practical terms, then, we were even more rural than the census figures show. Our institutions and traditions, and our Constitution, were

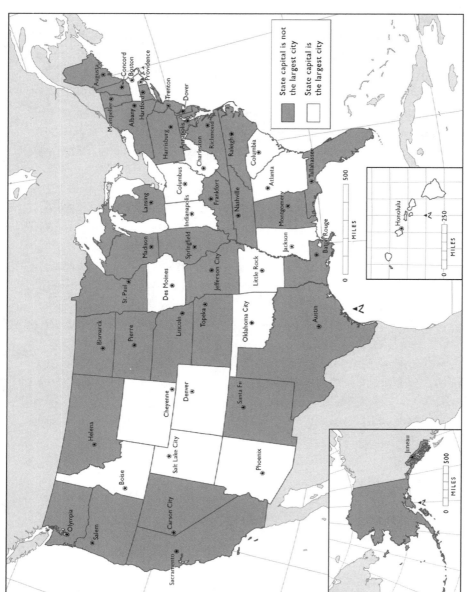

State Capitals of the United States

developed in the context of a situation where only a small percentage of Americans lived in villages or cities.

It wasn't until 1920 that America's urban population became as large as its rural numbers. Even well after that time, rural areas exerted more political force than urban areas in most parts of the country. There seems to have been a conscious decision to keep the center of state political power away from big cities. In some cases, a major cause was notoriously corrupt politics in the obvious big city choice; the selection of Sacramento rather than San Francisco for California is an example.

For whatever reason, the tendency to locate capitals away from the largest cities may have been a good idea, as we'll see when we later consider the question of large world cities.

Answer 6a: The most populated US state capital is, according to the 2010 census, Phoenix, Arizona. Around 7 percent of my students get that one correct. **Answer 6b:** As for the capital of Kentucky, it is, of course, pronounced "Frankfort."

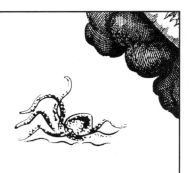

THE EXCEPTION TO EVERYTHING

Question 7: What county in the United States is well over a thousand miles in length?

There are over three thousand counties in the United States, although not every state has them (Louisiana has parishes, which are equivalent to counties). The idea of dividing territory into counties and having some degree of governance in those divisions comes to us from Europe and the British Isles, so counties have been around since before our Constitution was written. Hawai'i became a state only in 1959 and had ample time to learn from the experiences of the other states. One of the things it decided was not to have local government below the county level. There are no villages, towns, or boroughs in Hawai'i and only one city, but even that is called the City *and County* of Honolulu. The State of Hawai'i includes not only the inhabited islands in the chain but a long arc of uninhabited islands extending well into the North Pacific all the way to Kure **ATOLL**, more than a thousand miles away from Honolulu. **Answer 7**: These Northwest Hawaiian Islands are part of the City and County of Honolulu, thus creating the longest county in the United States.

Hawai'i truly is an exception to general rules about many things. Certainly in terms of climate, soils, geology, ethnic diversity, taxa-

Dr. G says: An atoll is a doughnut-shaped island made of coral. Its "hole" was where a volcano used to be, but as the volcano receded, living organisms called coral grew on the volcano's sides beneath the water. Coral, thought by many to be a rock or a plant, is actually an animal.

tion system (property taxes don't fund the public school system), official languages, land ownership, and customs, it's unlike the rest of the United States. In other parts of the world, people run away from erupting volcanoes, but in Hawai'i people flock to them. The relatively small volcano Kilauea has been erupting continuously for thirty years. Even the presence of volcanic islands in the middle of the Pacific is an exception. The so-called Ring of Fire, an area of frequent earthquakes and volcanic eruptions, is at the edge of the Pacific, not in the middle.

Land ownership in Hawai'i is more like the situation that prevails in Central America than on the mainland United States. About 90 percent of private land in Hawai'i is owned by a relatively few owners. The largest single private owner is the **KAMEHAMEHA** Schools/Bishop Estate. Revenue from these lands is used to fund a private school system that, with rare exception, admits only students who are of Hawaiian ancestry (which doesn't imply where the students were born or where they live but only their blood relationship to the aboriginal people of Hawai'i). Both the land policies and the admissions practices of the estate have come under the review of federal courts and seem likely to do so again.

Dr. G says: King Kamehameha united the Hawaiian Islands under his rule. The word *Kamehameha* means "the lonely one."

The state of Hawai'i recognizes two official languages: English and Hawaiian. The vocabulary of even English speakers in Hawai'i,

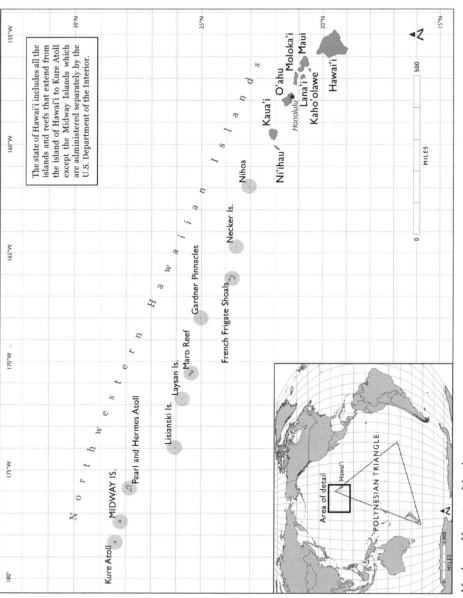

The state of Hawai'i includes all the islands and reefs that extend from the island of Hawai'i to Kure Atoll except the Midway Islands which are administered separately by the U.S. Department of the Interior.

Northwest Hawaiian Islands

however, includes a few hundred Hawaiian words, some of which have become increasingly used throughout the United States. *Mahimahi*, *ahi*, *muumuu*, and, of course, *aloha* seem to be universally understood. The most common language in Hawai'i, however, is a Creole language, incorrectly called "pidgin." Today it's based largely on English vocabulary, but because there are no Creole dictionaries and because it's not formally taught, it undergoes change from year to year and certainly from generation to generation.

The unique features of Hawai'i derive from one fundamental cause: its geographical isolation. Hawai'i is the most isolated inhabited land mass in the world and is about 2,500 miles from its nearest continental neighbor. The native species of Hawai'i, although clearly related to continental species, have evolved along different lines. The many ethnic groups that have come to make Hawai'i their home have contributed to the customs found there. The need for a common language has led to Creole, and the desire to preserve traditional foods has led to a situation where rice, rather than potatoes, is the common starch although the popularity of French fries may be a sign of change!

It's astonishing that 1,500 years ago the Polynesian people were able to sail from the South Pacific to settle Hawai'i. At that time, no other people possessed the knowledge and skill that would permit transoceanic voyages. The Polynesians used these abilities to settle a vast area, a triangle eight thousand miles on a side extending from New Zealand to Hawai'i to Rapa Nui (Easter Island). One of the great mysteries isn't why the Polynesians undertook such remarkable voyages but why they stopped doing it. Their navigational system depended on careful observation of the stars. My plausible and totally unproven assertion is that climate change produced much cloudier conditions and eventually the secrets of Polynesian navigation couldn't be passed on due to a lack of visibility needed to teach it.

MARRIED TO A STRANGER

Question 8a: What country is the leading trading partner of the United States?

Question 8b: From what foreign country does the United States import the most oil (petroleum)?

It seems that practically everything we wear and use in everyday life comes from China, and most everything we drive is made by an East Asian company. Americans are also sometimes shocked by the realization that the car rentals and airline reservations they make by telephone are with overseas agents, often in India. When it comes to oil, the media deluge us with the idea of our dependence on the Middle East. Nothing, it would seem, is done close to home anymore.

Answers 8a, 8b: My audiences are therefore naturally shocked and disbelieving when I tell them our biggest trading partner and our biggest source of oil is not in East Asia or the Middle East but rather that country north of most of us (and south of Detroit): Canada!

Even geography students studying for advanced degrees at American universities may study and specialize in a variety of world regions and know next to nothing about Canada. I supervised a PhD student who aspired to a career with the Foreign Service and who passed his written competitive exam with flying colors. Probably he could have spoken intelligently and well about the economies and political systems of any number of countries, but he was ambushed by questions in an interview about Canada. Most unfair, since the State Department generally seems even less interested in Canada than in Vanuatu!

The ancient Greeks developed a geographic concept called the "ecumene," which applies well to Canada. A country may cover extensive territory, but a much smaller, core territory is where most of the action happens. The Greeks, for example, would have considered the ecumene of Egypt to be the Nile River valley, and in Chile, the Vale (or central valley) would be the ecumene. In Canada, more than 90 percent of the population lives within a hundred miles of the border with the United States. Although Canada's ecumene doesn't cover the entire, lengthy border region, it certainly includes the upper St. Lawrence River Valley, south from Montreal and an extension through Ottawa to the north shore of Lake Ontario. We'd probably add in the rich farmland of the prairie provinces and urban Vancouver and Victoria on the west coast as well. That leaves an incredible amount of Canada that is non-ecumene: the vast Canadian Shield, ancient rocks with little agricultural potential, and the Arctic regions with no agricultural possibilities. What the ancient Greeks didn't understand was that areas that cannot sustain large populations may be important for another reason: natural resources. This is certainly true of Canada's non-ecumene.

While Americans may know little about Canada, Canadians know a great deal about the United States. Aside from their excellent educational systems, which feature some of the world's best university geography departments, the dominance of American entertainment, especially television, means that Canadians are deluged with information about the United States. Political geographers worry about this sort of situation and talk about "centripetal forces" that draw one country into the fabric of another. Canada, however, seems to have little trouble maintaining its strong separate identity.

Looking at the other side of the coin, television programs made in Canada for an American audience, while generally very good,

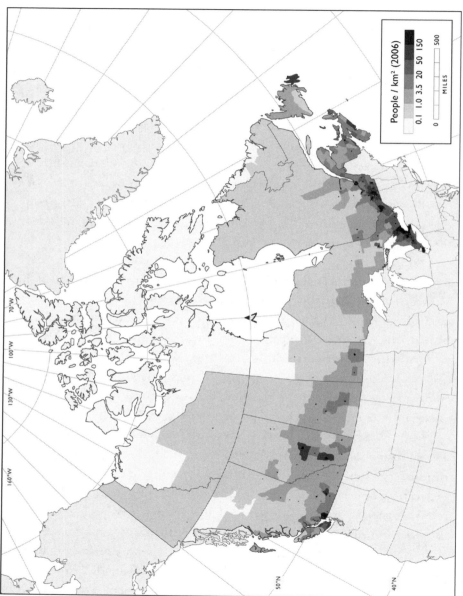

Canada and Its Population Distribution

People / km² (2006)

150
50
20
3.5
1.0
0.1

0   500

MILES

are, well, a bit strange. *Ice Road Truckers* is about as exciting as television can get, with maps showing the truckers' routes as they haul construction and drilling equipment to Canada's resource-rich non-ecumene. The audience, however, can't help but wonder about the mental health of anyone who undertakes such hazardous work: are all Canadians like that? Another exciting program, *Flashpoint*, is about a first-response police unit. Filmed in Toronto, it's replete with SkyDome, city scenes, and constables running about, but the fact that the setting is Canada is never mentioned in the program's stories. It almost seems that they pretend they're filming in a US city. The principal characters on the show, moreover, speak a variety of English that would be more at home in Toledo than Toronto.

WHY OLD MAPS
LOOK FUNNY

Question 9: What is longitude, and who first discovered how to accurately measure it?

The European sea powers that participated in the age of exploration and discovery were engaged in a highly hazardous undertaking, but of all the problems they faced none was more serious than not knowing where they were on the oceans of the world. Going aground on unknown islands was likely to be fatal, but smaller reefs and shoals were even more of a hazard. Sailing on a moonless night in unknown waters is scary business indeed!

A system for measuring distance north and south, latitude, was first developed by the ancient Greeks, then continually perfected by Muslim astronomers and geographers and further refined by the Portuguese. The instrument for measuring latitude was the astrolabe. So, for the early explorers of the Atlantic and Pacific oceans, measuring latitude was not a problem.

Answer 9: Measuring distance east and west, longitude, however, *was* a significant problem. Although theoretical systems were developed, they couldn't be used efficiently aboard ships. Maps and charts drawn during and after the early voyages of discovery look strange to us today. If you examine them carefully, however, you'll notice that

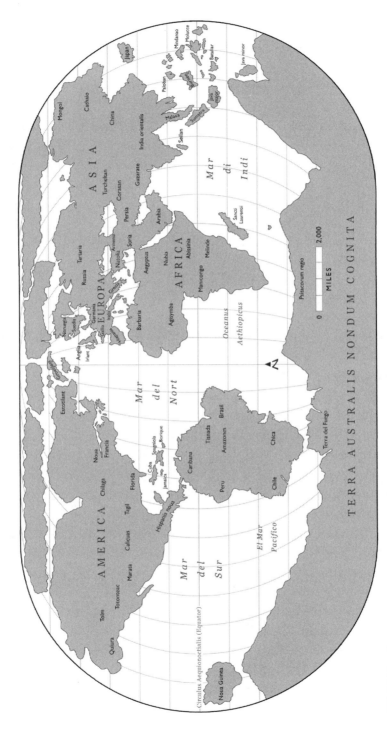

World Map from 1570

the north–south dimensions of land masses are reasonably accurate. It's the east–west direction that's distorted, due of course to the inability to measure longitude properly.

Numerous shipwrecks occurred well into the eighteenth century simply because longitude couldn't be measured. After several notorious shipwrecks, the British government offered £20,000 to anyone who could develop a system to measure longitude aboard ships. **Answer 9:** A British clockmaker, John Harrison, spent more than thirty years attempting to develop a timepiece that could be used by ships. The clock mechanism needed to survive the often violent movement of ships at sea. If time at the starting point of a voyage could be accurately known throughout the voyage, then it could be compared with local time and, after numerous computations, longitude determined. Harrison finally succeeded in 1761.

Harrison's invention gave the British a huge technological advantage over every other seagoing power, but the cost of Harrison's chronometer was so high that it was slow to come into widespread use. British ships continued to be wrecked by "longitude error" well into the nineteenth century.

In 1769, Lt. James Cook began his first voyage of discovery. His accomplishments on this journey included the rediscovery of New Zealand and the first European landing on the east coast of Australia. After landing in Botany Bay, he sailed through the Great Barrier Reef, a phenomenal feat of navigation and piloting. On his second voyage, beginning in 1772, Cook took four chronometers (none of them built by Harrison). One of them, called the K-1 chronometer, proved to be highly accurate and reliable, and it was identical to Harrison's invention.

Today, even the casual Costco customer can inexpensively navigate better than Captain Cook by simply purchasing a global positioning system (GPS). The GPS uses a system of linked satellites that enable drivers to avoid the use of maps. Unfortunately, geography textbooks are praising GPS technology as the greatest thing since sliced sourdough bread. But not all new technology is necessarily superior

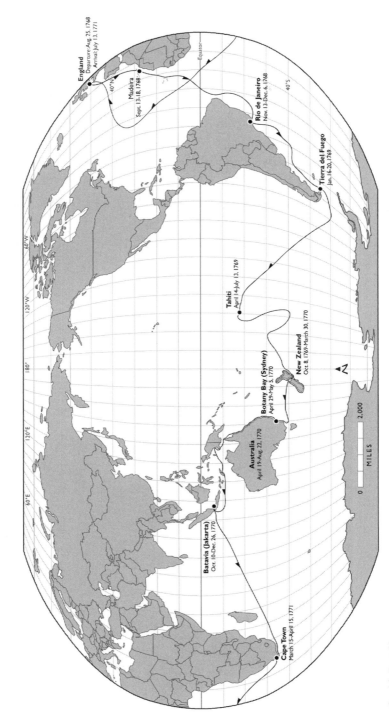

England
Departure: Aug. 25, 1768
Arrival: July 13, 1771

Madeira
Sept. 13–18, 1768

Rio de Janeiro
Nov. 13–Dec. 6, 1768

Tierra del Fuego
Jan. 16–20, 1769

Tahiti
April 14–July 13, 1769

New Zealand
Oct. 8, 1769–March 30, 1770

Botany Bay (Sydney)
April 29–May 5, 1770

Australia
April 19–Aug. 22, 1770

Batavia (Jakarta)
Oct. 10–Dec. 26, 1770

Cape Town
March 15–April 15, 1771

Equator

40°N

40°S

60°W

120°W

180°

120°E

60°E

N

0   2,000
MILES

Cook's First Voyage

to that which preceded it. At the Newark airport, in metropolitan New York City, staff were astonished when their GPS technology was shutting down about twice a day. They eventually discovered that truck drivers trying to avoid tolls on the nearby New Jersey Turnpike were using radio waves to block the signals that detected them when they passed through a toll plaza. GPS technology uses only a few watts of power to transmit signals from space and can thus be interfered with rather easily (not a comforting thought when aviation is involved). The older system, LORAN (LOng-RAnge Navigation), is land-based and likely more dependable than GPS.

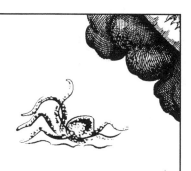

MILITARY USES OF GEOGRAPHY

Question 10: What is the world's highest national capital city?

Military applications of geography go back to prehistoric times. Understanding where the enemy is located and how the lay of the land makes certain points valuable for positioning troops are ancient uses of geographic principles. It's been said that around the fourth century BC, after the Greeks twice defeated Persian armies that invaded their territory, the Greeks wished to destroy the Persian Empire . . . but weren't exactly sure where it was! If this seems strange, recall that the modern United States has been repeatedly attacked by forces based somewhere in Southwest Asia. Exactly where these enemy forces are centered and where the United States should attack, however, remains in doubt, and the precise location of the enemy has been a major political issue in recent years.

In 1904, Halford Mackinder, a British geographer, published one of the most famous geopolitical articles of all time. He argued that there existed a geographic "pivot" in Eurasia, the control of which permitted control over the "world island" (Europe, Asia, and Africa), which in turn allowed control over the whole world. The German geopoliticians modified this idea somewhat with the slogan, "He who controls the Ukraine controls the heartland. He who controls the heartland controls the world island. He who controls the world island controls the world." At the time Mackinder wrote, the pivotal area of Eurasia was almost congruent with the Russian Empire, but Mackinder didn't believe that the Russians actually controlled the area militarily, and therefore they did not control the world. When Nazi Germany sought to impose Mackinder's ideas in 1941, the heartland they sought was no longer Russia but the Soviet Union. Operation Barbarossa, the invasion of the Soviet Union by Nazi Germany in June 1941, was the largest military plan ever undertaken.

Alfred Thayer Mahan provided a different perspective on geopolitics by advocating ideas that have been called "rimland control." Mahan believed that controlling the edge of the world island—in other words, sea power—was the key to world domination and critically important to the defense of the United States.

Mahan was an interesting personality. He was born at West Point, where his father was an instructor at the military academy. Over strong opposition from his family, he opted to attend the Naval Academy at Annapolis and graduated second in his class. He was remarkably unsuccessful as a deck officer and reputedly detested the steamships on which he served. His writings on sea power, however, made him world famous, and he came to the attention of President Theodore Roosevelt. Roosevelt cited Mahan's writings as justification for building the Panama Canal and for a two-ocean navy. Probably no military power embraced Mahan's ideas more than the Japanese, and their attack on Pearl Harbor on December 7, 1941, was a direct consequence.

Ironically, the Germans, who embraced the theories of Mackinder, also saw the merits of Mahan's rimland ideas. They attempted to take on the British fleet in World War I but were defeated decisively at the Battle of Jutland and never were able to get their fleet into the Atlantic.

Neither Mackinder nor Mahan anticipated the use of land-based air power, the carrier, or the submarine and the new dimensions

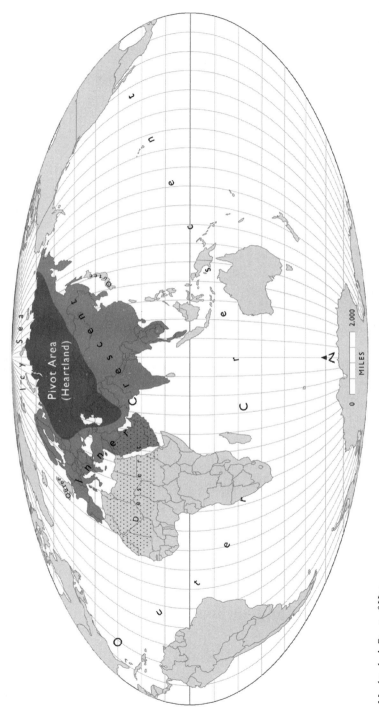

Mackinder's Pivot of History

they brought to warfare. Nevertheless, American geographers were in great demand during World War II, particularly in providing intelligence about the various battlefields in which US troops fought.

In more recent times, the involvement of American geographers with the military has declined right along with geography in our schools. One consequence of this occurred in 1969 when President Nixon appointed Nelson Rockefeller to head a commission to Latin America. Rockefeller was a poor choice since he wasn't popular in the region. Officials feared for his safety. The general security plan that was in place was a good one: in the event of hostile street riots, Rockefeller would be taken to his hotel, the entrance would be barricaded, and Rockefeller would be removed from the roof of his hotel by helicopter and taken to the airport.

It was never necessary to use the plan, which was fortunate in the case of one of Bolivia's capitals, La Paz. **Answer 10:** The helicopter to be used had a ceiling of only eight thousand feet, yet La Paz is the world's highest national capital at twelve thousand feet above sea level. It might have been handy to have a geographer on hand!

If you thought Lhasa, Tibet, was the world's highest national capital city, you need to think again. Lhasa *is* high, just a bit lower than La Paz, but the real problem is that Lhasa is not a national capital. Tibet is a part of China, and despite the "Free Tibet" bumper stickers you may see, not a single country recognizes Tibet as independent. (By the way, that's not a mistake in the preceding paragraph. Bolivia has two national capitals: La Paz and Sucre.)

Bolivia's Capitals

REALLY BIG CITIES

Question 11: What are the second largest cities in Mexico, Uruguay, and France?

The single greatest change that has occurred in the world over the last two hundred years may well be that we have increasingly become city dwellers. In 1800, only 3 percent of the world's population lived in places big enough to be called cities; now the figure is about 50 percent. The first city probably was Uruk in Sumer, founded about 3200 BC.

We began to develop really big cities after World War II. In 1950, the metropolitan area of New York City reached a population of 10 million, the first city to do so. Now there are eighteen of these "megacities." Because it's difficult to measure the precise population of megacities, several can lay claim to being the world's biggest. Different sources cite Tokyo, Mexico City, São Paulo, Shanghai, and Seoul as the largest. Another candidate might be what geographer Jean Gottmann called "Megalopolis," the continuous city extending from New York City to Washington, DC.

In 1939 geographer Mark Jefferson advanced the idea of the importance of the "primate city": the overwhelmingly largest city in a country. Sociologist George Zipf added some measurement to the idea and formulated the "rank size rule." Zipf noted that many countries, including the United States, had cities whose population size was related to the rank of its size. The second largest city was half the size of the largest city, the third largest was a third the size of the largest city, and so on down the whole rank order of all the cities in a country. The thirtieth largest city in the United States, for example, would have a population one-thirtieth the size of the largest, New York.

Zipf and others considered conformity to the rank size rule to be the normal state of affairs. If the rank size rule didn't apply to a country, then there was a basic flaw in its urban hierarchy. In particular, the largest city was likely to be much larger than all other cities. In practical terms, this meant that the primate city fed on its own growth: any urban function, whether it be manufacturing, educational institutions, service industries, government—most anything that goes to make up the infrastructure of a country—would be better off locating in the primate city. That's where there was a market for products, an adequate labor force, and transportation. Under these circumstances, the primate city would grow ever larger while growth would be restricted in other areas of the country.

Nonconformance to the rank size rule seemed to developmental economists to be often true of developing countries. Development schemes were often hatched to encourage growth in smaller, nonprimate, cities. The trouble with this approach was that (1) the evidence against the deleterious effects of primate cities was not entirely convincing, and (2) some highly developed countries also have primate cities. For us, however, a practical consequence of hugely primate cities is that other cities in the same country seem hardly to exist in our mental maps!

Mexico City, with 21 million people, is about five times larger than the second-ranking Mexican city, Guadalajara. At the same time, however, Paris, in highly developed, industrialized France, is about seven times larger than Lyon or Marseille. Don't bet huge sums on either of these answers, however. Guadalajara has only recently passed Monterrey in population size, and boundaries in the

Table 11.1. City Sizes in France, Mexico, and Uruguay by Rank and Approximate Metropolitan Population (Answer 11)

Country	Rank	City	Population (*in millions*)
France	1	Paris	10.2
	2	Marseille	1.4
	3	Lyon	1.4
	4	Lille	1.0
	5	Nice	0.95
Mexico	1	Mexico City	21.3
	2	Guadalajara	4.4
	3	Monterrey	4.0
	4	Puebla	2.7
	5	Tijuana	1.8
Uruguay	1	Montevideo*	1.3
	2	Salto	1.0
	3	Paysandú	0.08
	4	Las Piedras	0.07

*Including Ciudad de la Costa
Source: U.S. Census, International Data Base.

Mexico City area mean that some districts of metropolitan Mexico City may be larger than either Guadalajara or Monterrey. Meanwhile, although the metropolitan area of Lyon ranks second, the central city population of Marseille is larger than Lyon's. Be wary of trivia questions that ask about city sizes. As you can see, different answers to the same question can be "proven" by one source or another!

There's considerably more clarity when it comes to Uruguay. Montevideo is hugely primate and is more than thirteen times larger than the second largest city, Salto. If you answered this correctly, you're a true master of trivia!

Recall the earlier discussion on US capitals and you can see that locating those capitals in smaller cities has spread growth around considerably. As with the idea of primate cities in general, the jury is still out on the issue of concentrated versus dispersed urban growth.

AGRICULTURE AND TOURISM

Question 12: What country is the world's largest tourist destination, in terms of the number of visitors annually?

According to a 2005 news report, agriculture had ceased to be the world's leading occupation and was replaced by tourism! Whether that was true then (or is now), certainly the handwriting is on the wall. Farm workers continue to be replaced by mechanization everywhere, while only minimal replacement of workers by machines has occurred in the tourist industry. It still takes people to clean hotel rooms and ships' cabins. Tourism depends on quality service to survive, and service requires workers.

In some areas of the world, huge agricultural industries have all but disappeared and been replaced by tourism. Hawai'i and the Caribbean, for example, once big producers of sugarcane and other tropical crops, have become increasingly dependent on tourism. Condominiums for short-term rentals and part-time residents have become almost as numerous as hotel rooms and short-term tourists in these destinations.

Although employment in all kinds of agriculture has been declining worldwide, the type most often replaced by a tourist economy is "plantation agriculture." Plantations are a highly efficient way to produce crops and reached their zenith with sugar production. Plantation agriculture is monoculture, meaning it concentrates on a single crop. Huge tracts of land were devoted to it, and life centered on crop production. Housing was located near the fields, and social life was generally confined there as well.

Plantations are voracious consumers of labor. African slave labor was used on plantations extensively in the Western hemisphere, while elsewhere contract labor was employed. In Hawai'i, for example, plantation workers were brought from more than twenty countries but especially from China, Japan, and the Philippines. Because of the need to import agricultural workers, plantation agriculture is sometimes partly defined as the intrusion of one cultural group into another.

Tourism can be defined much the same way, but there's a difference. While plantation workers were expected to adopt some cultural norms of the host country, tourists aren't expected to do so in the countries they visit. It's increasingly difficult to find a tourist destination where English isn't the dominant language. For those who

Table 12.1. International Tourist Arrivals by Country of Destination, 2009

Rank	Country	Tourist Arrivals (*in millions*)
1	France	74.2
2	United States	54.9
3	Spain	52.2
4	China	50.9
5	Italy	43.2
6	United Kingdom	28.0
7	Turkey	25.5
8	Germany	24.2
9	Malaysia	23.6
10	Mexico	21.5

Source: UN World Tourism Organization.

Table 12.2. Employment in Agriculture as a Percentage of Total Employment by Country

Rank	Country	Percent
1	Rwanda	90.1
2	Laos	85.4
3	Tanzania	82.1
13	Cambodia	60.3
14	Albania	58.4
23	Morocco	47.0
24	Armenia	46.0
33	Guatemala	38.7
34	Uzbekistan	38.5
43	Egypt	29.9
44	Turkey	29.5
53	Brazil	21.0
54	Algeria	20.7
63	Panama	15.7
64	Costa Rica	15.2
73	Greece	12.4
74	Latvia	12.1
83	South Korea	7.9
84	Iceland	7.2
93	Spain	5.3
93	Estonia	5.3
117	Canada	2.7
126	United States	1.6
139	Macao	0.1

Source: World Development Indicators Database.

desire to learn what another culture is like, it's growing ever more difficult to do so as a tourist. Not only language but also accommodations and food are modified to serve the tourist.

Perhaps that's why the leading tourist destination is among those that least accommodate the culture of its visitors. **Answer 12:** France has more tourists annually than any other country, yet the shop signs are still in French and the worker in the *boulangerie* will still greet you with a *"Bonjour"* (and expects a reply *"en français, s'il vous plaît!"*). In one sense, it's hard to find a country that treats its visitors better than France; in another, the French seem to care very little about tourists and tourism. Maybe there's a lesson there!

(Many respond to the question at the start of the chapter with "Saudi Arabia," since they think of the requirement imposed on Muslims to visit the holy city of Mecca at least once in their lifetime if possible. Actually, the Islamic hajj is probably not even the largest of religious pilgrimages, which involves France again, if you consider visits to the shrine at Lourdes.)

CHAPTER 13

GEOGRAPHY AND RELIGION

Question 13: What is the only country in Europe with a majority Buddhist population?

For the past 2,500 years, the geography of the world has been changed dramatically by the expansion of the universal religions and the retreat of older, mostly **ANIMISTIC**, beliefs. Buddhism, Christianity, and Islam have changed the world. These religions seek converts—sometimes aggressively so, unlike earlier religions.

> Dr. G says: *Animism* is a general term referring to ancient religions that believed spirits were found in natural objects such as animals and plants, and even geographic features like mountains and lakes. Elements of animism can still be found in modern, universal religions.

The most recent universal religion to spread was Islam, whose followers are called Muslims. Islam began rapid expansion in the seventh century AD. It initially spread across North Africa and into areas now occupied by modern Spain and Portugal. This expansion was due to conquest by powerful armies from the Arabian peninsula originally led by the Prophet Mohammed, who was not only the founder of the faith but also a powerful and effective military leader.

Islam's spread to the east of its origin in Arabia was partly accomplished by missionaries who were primarily Arab traders carrying their faith along with their trade goods. By the fifteenth century, Islam extended from Portugal on the Atlantic Ocean to the southern Philippines in the Pacific. Islam was important throughout this huge expanse of territory and dominant everywhere except in India. Within the Islamic realm could be found great centers of learning and the most advanced practitioners of medicine, astronomy, and geography. In 1000 AD, an Islamic city, Baghdad, was the largest city in the world. Another Islamic city, Cordoba, in Spain, was the largest city in Europe. Today Islam is the most rapidly growing of the universal religions mostly because of the high birth rates of most Islamic groups.

Christianity preceded Islam and spread rapidly once it became the official religion of the Roman Empire in the fifth century AD. After being displaced in the Middle East by Islam, it became concentrated almost entirely on the European continent and in areas of Asia controlled by the Russian Empire. Because Europeans developed methods of transoceanic voyaging after 1500 Christianity spread overseas to North and South America, Australia, and New Zealand.

For most of the time Islam and Christianity have been in existence, they have generally maintained geographic separation. Historically, few Christians have lived in areas dominated by Muslims and relatively few Muslims have lived in areas dominated by Christians. Where there has been some mixing, or on the borderlands between the two religions, there has been significant conflict. Bosnia, Kosovo (territory disputed between Islamic Albania and Christian Serbia), Cyprus, Mindanao in the Philippines, and several areas of Africa are recent examples. Recently however, Muslim populations are growing in Europe. Turks in Germany, Indonesians in the Netherlands, Pakistanis in the United Kingdom, and North Africans in France are

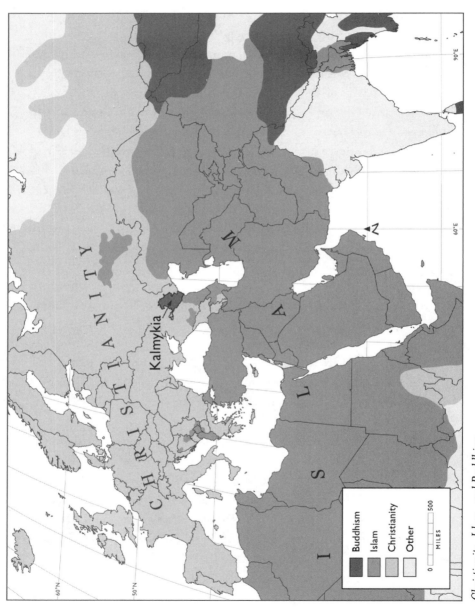

Christianity, Islam, and Buddhism

significant minority groups in areas where Muslim populations were once unknown.

Buddhism is the oldest of the universal religions and was founded in Northern India about 500 BC. While it actively seeks converts, some have disputed whether it should be called a religion. One of the major characteristics of a religion is that it involves worship. The Buddha ("the enlightened one") was asked directly by a disciple during his Deer Park sermons in Varanasi, India, "Whom do we worship?" The Buddha was silent.

Buddhism spread throughout South Asia, into Southeast Asia, and into East Asia as far north as Mongolia. Eventually it was absorbed by the older Hindu religion in India and became mixed with other religious beliefs in much of China, Korea, and Japan. One form of Buddhism, Lamaism, is found in both Tibet and Mongolia and is the state religion of Bhutan. **Answer 13**: Lamaism is also the majority religion of a European country: Kalmykia, an autonomous **EUROPEAN** country within the Russian Federation.

> Dr. G says: Europe has been defined for centuries as the area of the Eurasian continent west of the Ural Mountains in Russia. A highly vocal minority, however, insists that Europe is strictly a political entity containing only the members of the European Union. Such a definition would exclude all of Russia. It would also exclude Switzerland, thus leading us to wonder exactly where Switzerland is!

CHAPTER 14

GEOGRAPHIC POCKETS OF ISOLATION

PRESERVING OLD WAYS

Question 14a: *What is the oldest language in Europe?*

Question 14b: *Who was the first man to circumnavigate the world?*

Probably the most important feature of our contemporary world is change. Although some change has always been a feature of the world, the rate and magnitude of change under way now is unprecedented. The science of geography envisions change as a series of waves spreading over the surface of the earth. Each wave carries a new idea with it. Just like ocean waves, waves of change are affected by any number of things that change their size, shape, and speed. Ocean waves are stopped by land masses and waves of change are stopped by the isolation of places. Sometimes places are isolated by their distance from any other place, sometimes by harsh climate or other natural features, and sometimes by social isolation—that is, conditions where people resist change strongly.

You probably know people in your own neighborhood who attempt to isolate themselves, often to preserve a lifestyle that they've maintained and nurtured from the past. All people resist change to some extent, but some resist it very strongly indeed.

To understand this better, we need to look at extreme examples of isolation. Some Native American tribes in the headwaters of the Amazon are sufficiently isolated to have avoided most change. The Old Order Amish, a religious group once concentrated in Pennsylvania but now more dispersed, have rejected all new technology invented after the middle of the eighteenth century. Neither the Native Americans in Peru nor the Amish have been able to resist all waves of new ideas, but their isolation has kept them much as they were in the past. The location of the Native Americans has created their isolation, but the Amish have created their own social isolation.

Where we find isolation, we find less change and therefore a preservation of older ways. One of the more interesting isolated areas is in the borderlands between Spain and France. Although the area isn't particularly isolated today, it was for thousands of years. The **BASQUE** people, native to this region, resisted penetration by a variety of people who occupied adjacent lands. The Romans, who conquered and imposed their culture on most of France and Spain, weren't able to completely conquer the Basque people.

> Dr. G says: The language is usually called *Basque* in English, using the French word. In Spanish, the word is *Vasco*. In their own language, however, the language is called *Euskara*.

Answer 14a: Basque is considered by most linguists to be the oldest language in Europe. There are several ideas about its origin, but none has been proven. Most likely Basque was an ancient language, once spread widely in Europe and now preserved in the isolation of the Pyrenees Mountains.

In modern times, Basques have tended to migrate from their ancestral area, but some, after attaining success elsewhere, have returned to their homeland. Ironically, this once isolated region has become one of the more prosperous areas of Spain and has attracted outsiders

The Basque Region of Spain and France

seeking jobs. This has created great tension in the area, and a Basque terrorist group, ETA, has been responsible for hundreds of deaths.

Probably you've learned that Magellan was the first to sail around the world. Magellan's voyage was one of the greatest of all voyages of exploration, but he didn't actually complete it. He was killed in the Philippines. **Answer 14b:** The captain of the **ONE SHIP** in Magellan's original fleet to make it safely back to Spain after completing the circumnavigation was Juan Sebastian Elcano, a Basque!

> Dr. G says: Magellan began the voyage with five ships and 237 men. Only one ship, the *Victoria*, completed the voyage. Only eighteen men were left alive aboard the *Victoria*.

GEOGRAPHIC
INCOHERENCE

Question 15: Name a country whose national territory is completely divided by a foreign country.

In recent years, the term *failed state* has to come into use to describe a situation when a national government doesn't have effective control over all the territory in the country. An older term, used by geographers, is *incoherence*: when the parts of a country do not make a unified whole. Some incoherence is due to the lack of connectivity among parts of the country and some is due to simple geometry—the parts of a country don't fit together because they're separated by the territory of another country. Incoherence can be a serious problem for a country, which is why there have been notable attempts to turn incoherence into coherence.

From the 1840s until 1869, substantial populations of Americans lived in California and the Oregon Territory, but neither of these areas was well connected to the rest of the United States. This problem was partially solved with the opening of the transcontinental railroad when the **LAST SECTION** of track was laid with the Golden Spike put in place at Promontory Summit, Utah, on May 10, 1869.

Dr. G says: The railroad didn't become truly transcontinental until September 1869, when a bridge was completed over the San Joaquin River in California that connected the railroad to the Pacific Coast.

Similar situations existed in Canada and Russia. In both cases, railroads were built. In Canada, the trans-Canadian railroad connected British Columbia with eastern Canada. In Russia, the trans-Siberian railroad eventually connected Moscow with the Pacific port of Vladivostok.

Geometric incoherence is more difficult to solve, and if there is a solution it often involves direct military action, or at least the threat of it. Probably the most dramatic modern case involved the creation of the country of Bangladesh. In 1947, the colony known as British India was partitioned to form two independent countries: India and Pakistan. The new country of Pakistan was highly incoherent. Its eastern and western parts were divided by nearly one thousand miles of India. In 1971, the eastern part declared its independence and, while fighting went on for about nine months, eventually the new country of Bangladesh became fully independent of Pakistan.

Another case of incoherence, called the Polish Corridor, contributed to the start of World War II. In the Versailles peace treaties signed after the end of World War I, an old country was reborn in Europe: Poland. In 1920 **POLAND REEMERGED**, this time with a coastline on the Baltic Sea. The corridor that reached to the sea cut Germany into two parts. Germany never willingly accepted this split in its territory. World War II began in Europe with

Dr. G says: Poland had existed in the past but ceased to exist after new boundaries were drawn in Europe in 1815.

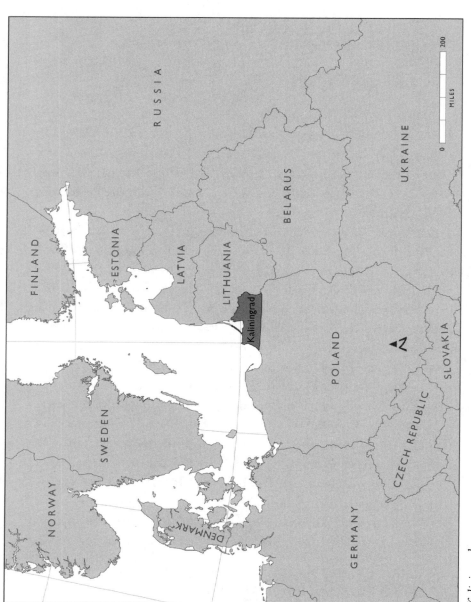

Kaliningrad

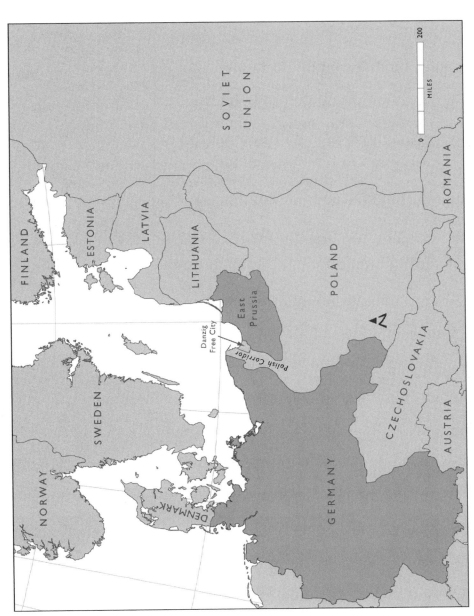

The Polish Corridor

Germany's invasion of Poland. The Germans quickly seized the corridor as well as the port city of Danzig (now Gdansk).

There are at least three examples of geometric incoherence in today's world. **Answer 15:** The Russian oblast of Kaliningrad is separated from the rest of Russia by Lithuania and Poland. Ironically, this territory surrounds the old German city of Konigsberg, the former capital of East Prussia, and it had been cut off from the rest of Germany by the Polish Corridor between 1920 and 1939!

Answer 15: The small, oil-rich country of Brunei, on the north coast of the island of Borneo, has its territory divided by a slice of neighboring Sarawak, a province of Malaysia.

Answer 15: Finally, an incoherent country overlooked by about 80 percent of my students is the United States! Alaska is separated from the rest of the country by Canada.

COUNTRIES
BREAKING APART

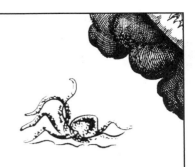

Question 16: What are the French speakers of Belgium called?

In the last decade several countries have threatened to break into pieces, and some actually have. The case best known to Americans is Canada. In recent years, elections have been held twice in the province of Quebec to see whether the people there wished to separate from Canada. Both times the vote failed by a small margin. If Quebec did leave Canada, Canada would be incoherent, with the Maritime provinces separated from the rest of the country. Opinion polls have shown that a solid majority of Quebec residents want change of some sort, whereas others in Canada are overwhelmingly opposed.

Canada is a bilingual country (French and English), but most French speakers live in Quebec. Language differences are certainly an important consideration in the issue of potential separation, but a broader concern is the feeling by the French-speaking Quebecois that they have not received a fair share of benefits and investments.

Meanwhile, the Soviet Union, Yugoslavia, and Czechoslovakia have already split apart. The general explanation for the breakup of these countries is nationalism—the desire of people who share a similar heritage (especially a common language) to govern themselves.

The force of nationalism is magnified when people who are living in their historic homeland are governed by people from outside.

Both Czechoslovakia and Yugoslavia resulted from the peace treaties after World War I. The Austro-Hungarian Empire that ruled substantial areas in central and Eastern Europe was dissolved and new countries were created based, in part, on the principle of nationalism. A famous geographer, Isaiah Bowman, was an adviser to President Woodrow Wilson and is often given credit for establishing the new countries that emerged from the treaty process.

In some cases, countries that represented a single nation were created. Poland and Hungary are two examples. In other cases, national groups were considered too small in number to warrant their own countries. The feeling by Bowman and others was that very small countries would lack vital natural resources and would fall prey to larger countries seeking territorial expansion. Czechoslovakia joined two national groups, Czechs and Slovaks, into a **SINGLE** country. Yugoslavia ("The Land of the South Slavs") became a stew pot of different peoples—Serbs, Croatians, Slovenes, and several others—in a single country. Unfortunately, some of these national groups had been ancient enemies, and Yugoslavia was bound eventually to become "the perfect storm" of ethnic conflict.

> Dr. G says: The former country of Czechoslovakia contained groups other than Czechs and Slovaks. The presence of a German population there gave Hitler's Third Reich justification to demand that a portion of Czechoslovakia be given to Germany. Eventually, Hitler took all of Czechoslovakia on the eve of World War II.

The Soviet Union was the direct heir to the Russian Empire, the world's largest country. Contained within its borders were dozens of national groups. After World War II, the Soviet Union added Latvia, Lithuania, and Estonia to its territory and further compounded the

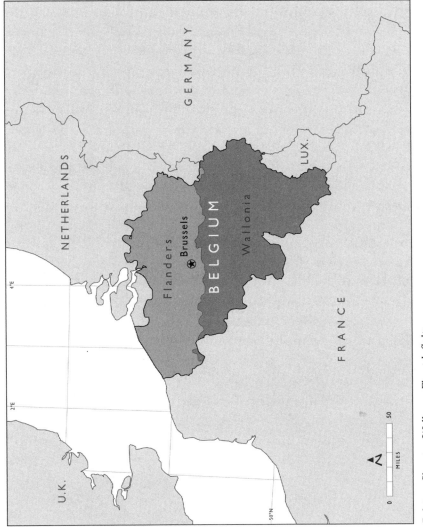

Belgium Showing Walloon–Flemish Split

problem of many nations within one country. So-called rising nationalism was the great fear of the Russian Empire, and this fear was
passed on to the Soviet Union.

Significant separatist movements occur in Scotland, Wales, the
Basque country, Catalonia in Spain, and several other countries
around the world. The most serious chance of fragmentation, however, is in Belgium. Two thousand years ago the area now occupied
by Belgium was the northernmost extent of the Roman Empire in
Europe. The southern half of Belgium, the part occupied by Rome,
became an area that spoke a Romance language while the northern half spoke a Germanic language. Although the north–south
division has blurred over time, today roughly half the population
speaks Flemish, a Teutonic (Germanic) language, while the other
half speaks a variety of French. Also like Canada, there's a religious
difference: the Flemish speakers are mostly Protestant, while the
French speakers are mostly Roman Catholic.

Answer 16: The French speakers of Belgium are called **WAL-
LOONS**. Walloon is actually a French dialect spoken by about 10
percent of the Belgian population, but because standard French is
more common, over time the name "Walloon" has come to refer to
all French speakers in the country.

Dr. G says: As was the case with many European groups,
some Walloons migrated to the United States. They
tended to settle in Wisconsin, particularly around Green
Bay and Sturgeon Bay.

UNITY IN DISUNITY

....

Question 17: In what European country are citizens required to keep guns in their homes?

....

These days it's common to hear of countries that have "failed" due to internal instability, including civil war, military coups, and insurgency groups, so it might be instructive to think about what attributes a country needs to be successful. Probably no single factor is more important than uniformity. If everyone in a country speaks the same language and practices the same religion, we might conclude that this is the kind of uniformity that makes for a successful state. While these unifying factors are certainly useful, they're neither sufficient nor necessary to make a successful state. Uniformity doesn't mean that everyone has to think alike, but it does mean that the people in a country have to feel some affinity for each other and agree that the country should exist. This basic idea is embodied in an idea called the "nation-state."

A nation is a group of people who have the necessary affinity—the needed uniformity—to provide the kind of stability a country needs for success. However, nations can exist without countries. There were, for example, both a German nation and a Polish nation before the modern countries of Germany and Poland came into existence. The land that a country occupies is also called a **STATE**. A perfect nation-state, then, is one where the boundaries of the state include *all* the members of the nation and *only* the members of that nation. Perfect nation-states, however, are rare. A number of states come close to perfection, but it's probably true that only Iceland qualifies as the perfect case: all Icelanders live in Iceland, and only Icelanders live in Iceland. Consequently, we'd expect Iceland to be among the world's most stable countries: no civil wars, no insurgency groups. Indeed, that's the case!

Dr. G says: Americans often get confused about the meaning of the word *state*, which usually refers to a country. The word is also used to refer to the fifty units of the United States because, following the American Revolution, the thirteen colonies became thirteen independent countries prior to the Articles of Confederation and the Constitution.

Every country other than Iceland, then, is a less than perfect nation-state. For a variety of reasons (in these other states), it's been necessary to compromise with perfection. Now let's turn the coin over and try to design the worst possible nation-state, as unlike Iceland as we can imagine. Our worst case would be a state with more than one nation. Moreover, we'd have each nation live on land it considered to be its heritage, its historic homeland. We'd allow very little mixing; each nation would have its own corner of the state. To make it even worse, each nation would have its own language; there'd be no overall common language in the country. Next we'd really make it messy. Most of the nations living in our hypothetical state would have their own "Mother" nation-states *just across the border*! This geographic situation implies that the big nation-states next door might (and probably would) try to expand their borders to absorb their nationals. Our highly imperfect nation-state would thus have great internal diversity and the constant threat of invasion by its neighbors. In such a case, we'd expect our imperfect creation to be the ultimate failed state.

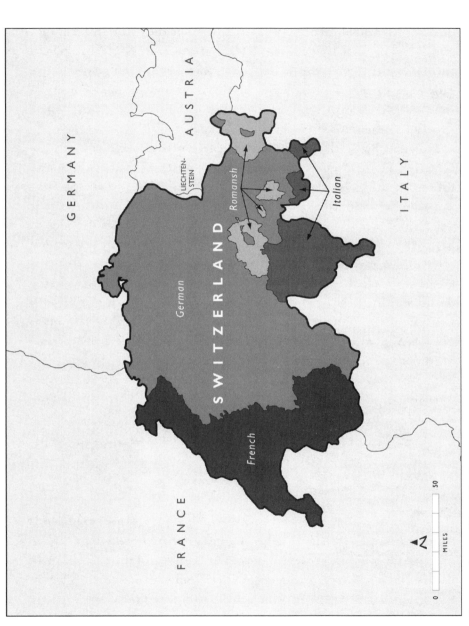

Linguistic Areas of Switzerland

The "most imperfect" nation-state we have created turns out to be Switzerland! This country has existed since the thirteenth century and in its present form since 1848. Different areas of Switzerland speak German, French, Italian, and Romansh (a language found only in Switzerland). It's one of the richest and most stable countries in the world. Considering how "imperfect" it is, Switzerland is just short of miraculous.

How has Switzerland achieved perfection with such lack of uniformity? There are probably three major unifying factors. First and foremost, the Swiss identify with Switzerland. There's a strong feeling of being Swiss rather than German, French, or Italian. In other words, the idea of belonging to a Swiss nation overrides the idea of being French, German, or Italian. Second, Switzerland is a federal state with a great deal of local autonomy. Governance tends to be at the municipal, or **CANTON**, levels so that feelings of being dominated by other national or linguistic groups are minimized. Uniformity as measured by language exists at the local level even as it's absent at the country level. Third, the Swiss have a deep commitment to the unifying principle of neutrality. They stayed out of both World War I and World War II despite the fact that both wars were fought all around them.

> Dr. G says: Swiss cantons are somewhat equivalent to a province, but because of their small size they're often considered to be like British or American counties.

Because it's such a peaceful country, the rest of the world tends to forget that Switzerland helps maintain its neutrality by a strong military force. Men between the ages of twenty and thirty must serve in the Swiss military if they're physically able. **Answer 17**: They're issued automatic weapons, which they're required to keep with them at home. When their period of military service ends, they may keep their firearms, but the weapons have to be rendered incapable of continuous (automatic) firing.

MAPS FOOL US AGAIN!

Question 18: Of the forty-eight contiguous United States, which is the most northerly?

Each year since 1989, a **NATIONAL GEOGRAPHY BEE** has been held, with the finals in Washington, DC. One year, according to reports, a major controversy developed over this question.

Dr. G says: The event was originally called the National Geography Bee, but the name was changed to the National Geographic Bee.

If the question were asked about the fifty United States, the answer would obviously be Alaska, but when we ask about the forty-eight contiguous states, most people visualize the US–Canadian border as a curved line extending from Washington State to Lake Superior and continuing east to Maine. Given this perspective, Washington State becomes a likely answer since it appears to be at the north end of the curved boundary line. The other choice is Maine, which also looks pretty far north. About 94 percent of my college students picked either Maine or Washington State when asked this question.

In reality, that "curved" appearance of the northern boundary of the United States is an artifact of map projection. The line is actually straight; Washington is not north of Idaho, Montana, North Dakota, or Minnesota. Maine is south of these five states, so it can't be the right answer, either.

Here's where the geography bee controversy came in: apparently, the issue of the US–Canadian border was discussed in materials sent out before the contest. It was stated in the materials that Isle Royale, in Lake Superior and a part of Michigan, was the most northern part of the forty-eight states. When asked this question during the actual contest, a contestant answered "Michigan" . . . and was declared wrong!

Answer 18: The correct answer is Minnesota. An area referred to as the Northwest Angle and containing the town of Angle Inlet protrudes slightly north of the rest of the US–Canadian border. The Northwest Angle is on the shores of Lake of the Woods, the sixth largest freshwater lake in North America (after the five Great Lakes), and contains 65,000 miles of shoreline. Although the Northwest Angle can be reached by land by travelling through Canada, it's reachable from the rest of Minnesota only by water.

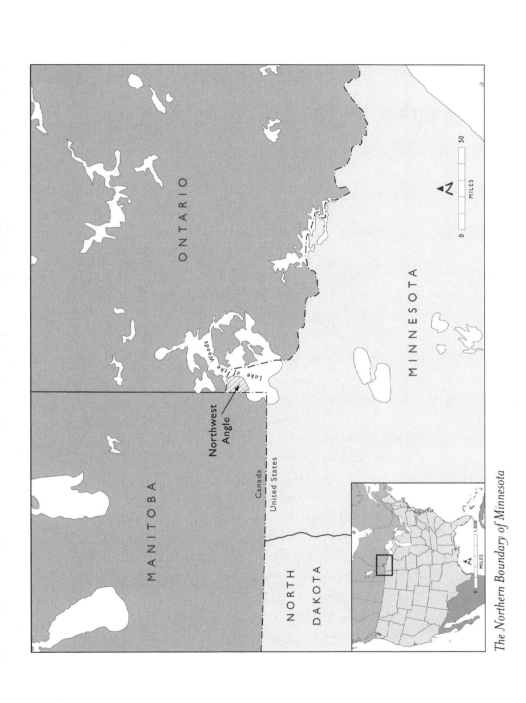

The Northern Boundary of Minnesota

CULTURE HEARTHS

ANCIENT AND MODERN

....

Question 19: What do Mormonism, baseball, and women's rights have in common?

....

Culture is that super important word that describes the learned behavior that we pass on from generation to generation. It includes language, belief systems (including religious belief), art, music, tools, technology, and a host of other things that separate "our" group from another group. We used to believe that physical differences, which we called "race," were what separated us into groups. We've learned that physical differences are minor; it's learned behavior that both unites those that share the culture and separates those groups that don't share it.

Every facet of culture has to start someplace, and if a place has served as a source of new cultural features for a long period, we call it a "culture hearth." Mankind's most important culture hearth is probably Southwest Asia. Most of the domesticated plants and animals we use originated there, as did many agricultural tools and farming systems. Religious beliefs—including Judaism, Christianity, and Islam—came from this area as well.

Other areas also served as significant culture hearths. Thanks to the work of geographer Carl Sauer and anthropologists, we now know that a hearth in Southeast Asia was more important than originally thought. Sauer argued that the earliest agriculture actually began in Southeast Asia and was based on the planting of shoots and cuttings rather than seeds. A hearth in **MESOAMERICA** produced only a few agricultural crops, but one in particular proved to be exceptionally valuable. Maize, or corn, required an incredible amount of crossbreeding before the modern crop was developed. Now, in addition to being the staple grain in many parts of **LATIN AMERICA**, sugar and vegetable oil are also made from corn.

Dr. G says: The terms *Latin America* and *Mesoamerica* are confusing. While *Latin America* should refer to those countries in the Western hemisphere that speak a Latin-based language, in practice the term refers to all those areas south of the southern US border, regardless of the languages spoken there. *Mesoamerica* ought to refer to "middle America," Mexico and Central America. In common practice, however, it often also includes the Caribbean islands.

The first culture hearth in the United States probably developed in the early part of the nineteenth century. The original thirteen colonies, and the states that were derived from them, were basically European in terms of their culture. Even our political system, which was unique, was clearly derived directly from British and French ideas. How did we develop a culture that was distinctly American?

At the close of the American Revolution, areas to the west of the Mohawk Valley in New York, formerly controlled by tribes loyal to the British, were opened up for settlement. The land was good for agriculture, and settlers flocked there from all parts of the United States. It was the first occasion when subcultures from various parts of the new United States commingled. The rich soil deposited by melting glaciers, called "till," provided the means to both produce a surplus and attract diverse people.

Between 1820 and 1850, a number of new ideas, perhaps showing European ancestry but decidedly American, began to emerge
from this region. Baseball was one of these new ideas. The National
League, after an investigation, concluded that Abner Doubleday
invented the sport around 1839. Doubleday lived in Auburn and
Cooperstown, New York, where the Baseball Hall of Fame is located.
Recently, some historians have challenged the idea that Doubleday
invented baseball, but there has been no challenge to the idea that
baseball developed in this area of central and western New York.

Several religious movements also are associated with this first
American hearth. Perhaps most typically American is the Church
of Jesus Christ of Latter-day Saints, the Mormons. This religious
movement was started by Joseph Smith in the village of Palmyra,
New York, around 1820. The Mormons moved west in the face of
strong discrimination, eventually settling in Salt Lake City, Utah.
An annual pageant is held on Hill Cumorah, just outside Palmyra,
celebrating the discovery of golden tablets by Smith that marked the
beginning of the movement.

The women's rights movement and the **ABOLITIONIST**
movement were either born in this hearth area or later became
strongly associated with it. Elizabeth Cady Stanton is credited with
beginning the women's rights and **SUFFRAGE** movements in the
United States. She was born in Johnstown, New York, and the first
women's rights convention was held in Seneca Falls, New York, in

Dr. G says: The abolitionist movement advocated for the
abolition of slavery in the United States, and the suffrage
movement sought voting rights for women. Although
both movements were closely linked before the Civil
War, they split after the war over the ratification of the
Fifteenth Amendment to the Constitution. That amendment granted former male slaves the right to vote but
didn't offer the same right to women.

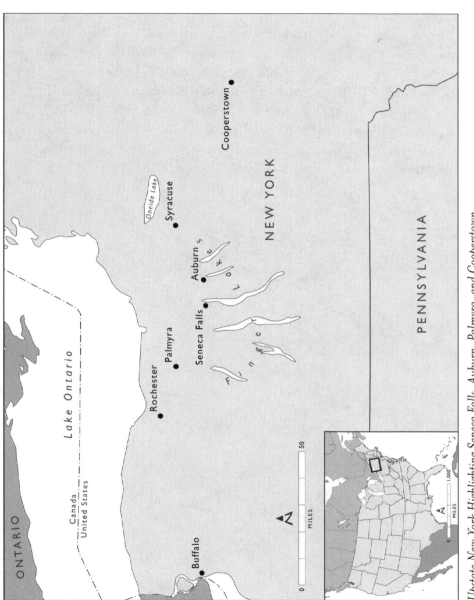

Upstate New York Highlighting Seneca Falls, Auburn, Palmyra, and Cooperstown

1848. William Seward, secretary of state in both Lincoln's and Andrew Johnson's cabinets, was one of the strongest voices for the abolition of slavery. From his home in Auburn, New York, both Seward and his wife were leading social and political activists in their day.

To my knowledge, there has been no complete inventory of the social, political, religious, and technological ideas that emerged in the western New York hearth between 1820 and 1850, but the list must be impressive. What do baseball, the women's rights movement, and Mormonism have in common? **Answer 19**: They all developed in a small area of western New York State between 1820 and 1850.

POPULATION DECLINE

CHANGE ON THE HORIZON

Question 20: What country has fewer people today than it did 160 years ago?

In the 1960s, issues concerning world population were center stage. Some authorities believed human population problems were the greatest threat to human survival, more so even than nuclear warfare. The precise nature of the threat was unclear. The word *overpopulation* was bandied about, and many people thought they knew what it meant, but no consistent definition was advanced. To some, the threat was the sheer number of people on earth, to others it was the rate of population growth. Some argued that there was a limit to the earth's "carrying capacity," while others talked about "overcrowding." Ironically, there are now more than twice as many people on earth as when the debate raged, yet issues concerning world population have receded. Why has concern faded?

Around 1962, the world population growth rate peaked at about 2.2 percent annually. It has been slowly dropping ever since. Currently the growth rate is estimated to be 1.1 percent. While this is still high by historical standards (over the last ten thousand years, the average rate of growth has been 0.1 percent or less), the fact that it's only half of what it was fifty years ago helps explain why there's less concern today.

Forecasting future population numbers is not easy, but if we make some reasonable assumptions about birth and death rates, we can expect world population to reach its maximum between 2040 and 2100 and begin to decline. At that point, total world population should be between 8.1 and 10 billion. The estimated world population for 2011 is 7 billion.

Despite the continued overall growth of world population, some countries will face actual population decline in the near future, and some are already in decline! One of the largest declines, both in percentage terms and actual numbers, is in Russia. According to UN estimates, Russia's population will decline by at least 20 million by 2025. For every 100 births in Russia, there are 160 deaths, and it's all but certain that this disparity between births and deaths will increase in the future. Several other countries (some of them parts of the former Soviet Union) are also in decline.

In addition to the countries already showing population decline, a number of others have fertility rates that are "below replacement." This means that unless there's a future increase in the birth rates in these countries, they, too, will begin to decline in population (however, in some of these countries migration may replace the losses). For the most part, these are European countries, although Japan is on the list as well.

In Table 20.1 I use a measure of human reproduction called the total fertility rate (TFR), which indicates the number of children that will be born, on average, to each woman over her entire reproductive lifetime. A woman and her mate require two children to replace themselves, but because some women cannot bear children, it requires an average of **2 . 2 CHILDREN** to replace the existing population. Therefore, whenever the TFR falls below 2.2 and remains there, total population will decline eventually. Currently in some countries, immigration is compensating (or partially compensating) for the declining total fertility rate, but very dramatic change (in either birth rates or immigration) would be needed to stave off overall population decline.

Table 20.1. Countries with Fertility Rates Substantially Below Replacement

Rank	Country	TFR[a]
1	Denmark	1.74
2	Finland	1.73
3	Trinidad and Tobago	1.72
4	British Virgin Islands	1.71
5	Tunisia	1.71
6	Barbados	1.68
7	Sweden	1.67
8	Netherlands	1.66
9	Belgium	1.65
10	Thailand	1.65
11	Cuba	1.61
12	Canada	1.58
13	Macedonia	1.58
14	China	1.54
15	Portugal	1.50
16	Albania	1.49
17	Spain	1.47
18	Switzerland	1.46
19	Georgia	1.44
20	Austria	1.33
21	Serbia	1.39
22	Slovenia	1.29
23	Bosnia and Herzegovina	1.26
24	Czech Republic	1.25
25	South Korea	1.23
26	Taiwan	1.10

Source: Author's estimates based on data from the International Data Base (US Census, Suitland, MD).

[a] TFR = total fertility rate. Countries with a TFR below 1.75 will likely begin to decline in population by 2020.

Table 20.2. Countries with Current Declining Populations and Estimated Percentage Decrease by 2050

Rank	Country	Percent Decrease by 2050
1	Bulgaria	35
2	Romania	30
3	Ukraine	29
4	Estonia	24
5	Latvia	23
6	Russia	21
7	Japan	20
8	Poland	18
9	Lithuania	15
10	Croatia	14
11	Moldova	14
12	Belarus	13
13	Slovakia	12
14	Hungary	12
15	Germany	10
16	Slovenia	7
17	Italy	6
18	Greece	3

Source: Author's estimates based on UN and World Bank Data.

Dr. G says: In developed countries like the United States and Canada, the total fertility rate required to replace the population is about 2.1, while in lesser developed areas, higher mortality may require a TFR of 2.3 or higher for replacement.

In brief, the future population problem will be one of decline, not increase. This presents unprecedented problems to the affected countries. One of the most pressing will be Social Security or its equivalent. This means of supporting retired people is based on the notion that there will be many working people and a relatively few retired. This has now flip-flopped, and paying retiree pensions is becoming increasingly difficult.

Ireland (both the **REPUBLIC OF IRELAND** and **NORTHERN IRELAND**) is an interesting exception. These territories reached a historic low point of population in the 1960s of around 4.2 million but have had a population boom since then with a current population of 6.2 million. **Answer 20:** This is especially ironic because Ireland may be the only country in the world to have fewer people than it did 160 years ago (when the population was over 8 million).

Dr. G says: The entire island of Ireland was a division of the United Kingdom in 1840 when it reached its maximum population. In 1922, the island split into the Irish Free State and Northern Ireland (which remained part of the United Kingdom). The Free State became the Republic of Ireland in 1949.

GEOGRAPHIC DIFFERENCES AND CIVIL WAR

Question 21: What US state was named after Queen Elizabeth I?

The magnitude of the US Civil War stuns modern Americans. More soldiers were killed in the Civil War than in all other wars combined in which the United States has fought. Far more died in a single day of battle than have been killed in all the years we've been engaged in Iraq and Afghanistan. The first battle of the Civil War, called First Bull Run (by the Union) or First Manassas (by the Confederates), was the largest battle ever fought in the Western Hemisphere up to that time, yet it was small compared to other battles fought later in the Civil War.

Debates among historians about the causes of the Civil War have gone on ever since the war began. Some argue that slavery in the South (and its near absence in the North) was the main cause. Others cite the economic differences: the South depended almost entirely on agriculture, while the North had a much more diverse economy. Some have even argued energy-use differences: the South depended heavily on human and animal muscle power, while North- erners used water power to run their mills, and they used more coal.

I can't settle the debate here, but I can offer the idea that a geo- graphical dimension to the arguments is often overlooked. Perhaps the most important forgotten geographic feature is the direction of the mountains, the Appalachians, which extend from Pennsylvania to Alabama. These mountains trend from southwest to northeast. A Confederate army that followed the Blue Ridge Valley and other valleys trending northeast would find themselves in the heart of the North, within easy striking distance of Philadelphia and New York. A Union army following the same valleys to the southeast would end up in **DOGPATCH**! In other words, even before the Civil War started, the South held a strategic geographic advantage considering only land-based warfare.

> Dr. G says: Dogpatch was a parody of a southern mountain community in Al Capp's comic strip, *L'il Abner*.

Also overlooked are the lifestyle differences that existed between the North and the South that go beyond economic differences. The South was overwhelmingly rural at the time of the Civil War, and the **SEVEN SOUTHERN STATES** that seceded from the United States before the Civil War started were especially rural. Plantation agriculture and isolated small farms were the predominant settlement pattern. While the North also had a substantial rural population, the rural countryside was punctuated with villages. Life there tended to center around the villages and small towns. Even today, when great changes have transformed both north and south, it's still possible to claim that the biggest difference between the regions is a rural versus urban outlook.

> Dr. G says: The first seven states to secede were South Carolina, Mississippi, Alabama, Georgia, Florida, Louisiana, and Texas.

Long before the Civil War, the southern colonies tended to be aligned closely with the Royalist interests in Great Britain. During

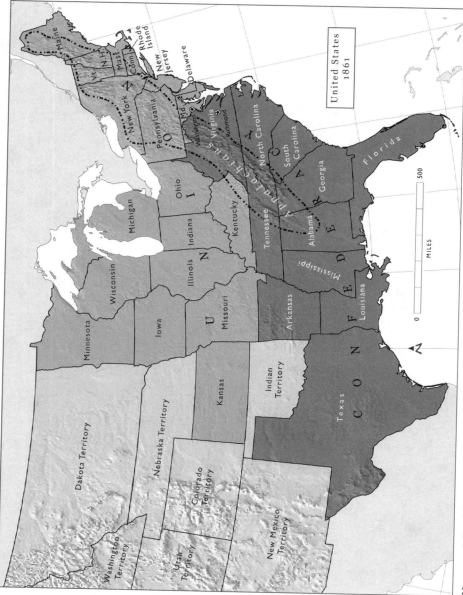

Union and Confederate States

the American Revolution, the British recognized that the "rebels" were politically strongest in New England. Their "three pronged" military campaign of 1777 attempted to split New England from the rest of the colonies, but none of their "prongs" was successful. General William Howe was unsuccessful in moving his troops north on the **HUDSON RIVER**, General John Burgoyne lost the Battle of Saratoga (New York), and Colonel Barry St. Leger was unable to take Fort Stanwix (near present-day Rome, New York) and tried to join up with Burgoyne's army but arrived too late. Following these defeats, the British focused their military activity in the South, where Royalists (Tories) were more common.

> Dr. G says: A huge chain had been placed across the Hudson at West Point to prevent a British fleet from sailing north to Albany. Later in the Revolutionary War, the infamous American traitor Benedict Arnold was involved in an unsuccessful plot to turn West Point over to British forces.

This difference in political outlook can still be seen on a map today. **Answer 21:** None of the Northern colonies were named after a British monarch (Massachusetts, New Hampshire, Connecticut, Rhode Island, New York, New Jersey, Pennsylvania, Delaware, and **MARYLAND**), but all the Southern colonies were named after a king or queen (Georgia after King George; North and South Carolina after King Charles; and Virginia after the "virgin queen," Elizabeth I).

> Dr. G says: While Maryland wasn't named after a British monarch, it was named after the wife of Charles I, Henrietta Maria, who held the title of "Queen Consort."

NEW AGRICULTURAL CROPS

....

Question 22: What's the only edible orchid, and where is it commercially grown in the United States?

....

Make no mistake, the most valuable agricultural crops produced today, either in total value or in price per acre of produce, are illegal crops. The huge and seemingly insatiable demand for recreational drugs made from poppies, coca, and marijuana has created a huge market and enormous wealth for those who are successful in the trade. This in turn has created a nightmare with consequences that reach into all our lives.

Due in major part to continued technological advances in farming, much agricultural land in the United States has been replaced by housing or simply been abandoned. There have been, however, some interesting new developments in the use of farmland. Relatively new legal crops have emerged, and existing (but unusual) crops have seen an expansion in production.

As Americans have become more accustomed to dining out, a demand has been created for specialized crops used to make salads. So-called micro-greens have become a valuable crop, usually grown on relatively small plots of land close to urban areas. A large range of **ORGANIC** fruits and vegetables have begun to appear on restaurant menus and supermarket shelves. Like micro-greens, organic crops, in general, command a high price and offer a reasonable economic alternative to farmland becoming housing developments.

Dr. G says: Organic crops are grown using a minimum of synthetic fertilizers and pesticides. The practice of organic agriculture requires government certification.

Jack-o'-lantern pumpkins have become a valuable seasonal crop in near-urban farmlands. Many growers don't harvest the pumpkins themselves but sell directly to consumers who visit farms and select their pumpkins from the field. Other crops, such as strawberries, oranges, and grapefruit, also are sold on a pick-your-own basis, thus saving on labor costs.

The most valuable crop in the world, by weight, is saffron. It takes approximately 225,000 stigmas from the saffron crocus to equal one pound of saffron, but that pound can sell for as much as $5,000! The Amish in Lancaster County, Pennsylvania, cultivated saffron well before the American Revolution, and today Pakistan, India, and Spain are the leading producers. Because of the manual labor needed, it's grown commercially in the United States only in small amounts.

Vanilla is the world's second most valuable crop (by weight or value per acre). It's native to Mexico, but the big producer is the island country of Madagascar, located off the east coast of Africa. One often hears about "vanilla beans," so it's natural to assume that vanilla is cultivated like a bean. **Answer 22:** In fact, vanilla is an orchid, the only edible one! The vanilla orchid blooms after about five years of growth. The blooms last for only a few hours, and during that time, the blossoms have to be pollinated by hand. If the pollination is successful, then the bloom is replaced by pods that resemble string beans.

Although vanilla is cultivated by hobbyists in semitropical locations in the United States, the only commercial vanilla farm as of this writing is on the island of Hawai'i, near the small village of Pa'auilo. The Hawaiian Vanilla Company welcomes visitors and serves a gourmet lunch in which all the dishes contain vanilla.

DOMESTICATIONS

Question 23a: For what animal were the Canary Islands named?

Question 23b: What's the geographic difference between a yam and a sweet potato?

Question 23c: What animals did the Native Americans domesticate?

The domestication of plants and animals, collectively known as the Agricultural Revolution, is arguably man's greatest accomplishment. Animal domestication is perhaps more significant because it involves risk. Resowing wheat that stays longer on the stalk and is therefore easier to harvest doesn't involve much danger, but what about wolves? Evidence indicates that the wolf was the first domesticated animal and is the ancestor of today's dog. Stories of the dangers of wolves are common, even in children's literature, yet the collective affection we have for dogs—and our dependence on them—has more than justified the risk that our ancestors took in this first domestication step.

When Europeans first encountered the aboriginal people of the Canary Islands, they observed people with a limited material culture and technology. So limited, in fact, that they didn't appear to have had the means to have gotten to the islands! Their presence is therefore a mystery, but they did have a domesticated animal. **Answer**

23a: The Canary Islands were named after an animal, but probably not the one you had in mind. The name comes from the Latin word *canis*, for "dog," probably a reference to feral dogs that were noted in the Canary Islands by early European voyagers. At least one account of the origin of the name suggests that it may have come from a now-extinct colony of seals that once inhabited the Canary Islands. Since seals "bark," it's suggested that the islands were named for the sound dogs make rather than for dogs themselves. In the very unlikely event that you answered "seal," take credit for a right answer!

The Agricultural Revolution, despite its importance, is often misunderstood. My fourth-grade "citizenship education" textbook contained a story of a cave dweller who accidentally spilled some seeds from plants he had gathered for dinner. The following spring, the seeds sprouted where he had spilled them . . . and Kowabunga! . . . agricultural revolution. Such a chance discovery can hardly qualify as one of mankind's greatest accomplishments, and my textbook failed miserably in portraying the near miracle involved. The revolution in agriculture meant that the same amount of land, water, sunshine, and labor could produce more food. If it's not the proverbial free lunch, it's the closest thing to it.

The Agricultural Revolution involved the domestication of plants and animals. By selective breeding, man has been able to continually **IMPROVE** a number of plant and animal species to enable more food production, better transportation, and even more advanced warfare. So completely have we transformed some species that they bear only a superficial resemblance to what we started to domesticate. The Chihuahua, for example, doesn't seem much like a **WOLF**! Botanists are astounded at the path that selective breeding has taken in the case of corn, or maize. Mesoamerican Indians made maize a staple grain in their civilizations, but once it became an important crop for colonial American settlers and later began to be cultivated globally, change occurred even more rapidly.

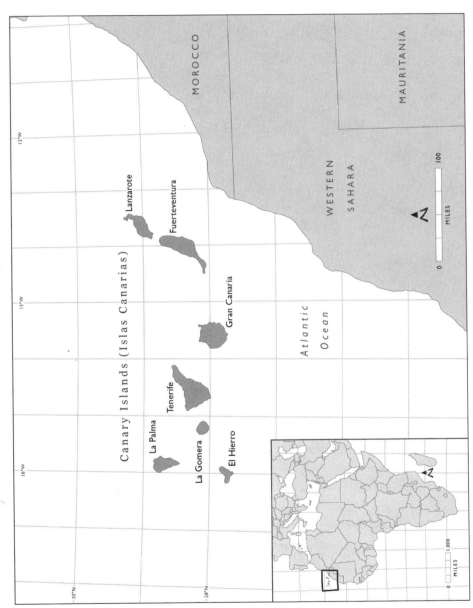

The Canary Islands

Dr. G says: Whether a man-induced change in a species is actually an improvement is a difficult and sometimes debatable question. Certain fruits and vegetables, for example, have undergone change to allow their transport over long distances without spoilage. This improves the availability of food but not necessarily its flavor or its nutritional value. Currently, genetic modification of crops has enormously speeded up the change in species and generated much controversy. In animal species, the modifications in the English bulldog have created a major controversy among dog breeders and pet owners in general.

Dr. G says: There's been a great deal of debate about the origin of dogs, but DNA evidence has established that all dogs are descended from wolves, albeit different packs of wolves in different parts of the world.

Since the domestication of plants and animals occurred over thousands of years and still continues today, it doesn't seem proper to call it a "revolution." Still, the process, more than anything else, has made man the dominant species on the planet . . . and that's pretty revolutionary! Agricultural revolution provided a more reliable source of food. Food security is certainly not universal: hunger and starvation are still present, and famines still occur. For the most part, however, the world is better fed than at any time in history. Parents of my generation used to remind us to "eat your spinach because children are starving in India." While the link between spinach and hunger in Asia was never entirely clear to us, it seems astonishing to realize that India now exports more food than it imports.

Hearth areas, mentioned in an earlier chapter, are linked primarily to the plants and animals they domesticated. The most important hearth, Southwest Asia, gave us wheat, barley, flax, and dates, while Southeast Asia contributed rice, bananas, and citrus. The Meso-american hearth is where maize, sweet potatoes, squash, and tobacco originated, while South America gave us the potato.

There are some mysteries and confusion concerning the origin of certain agricultural crops. Peanuts, for example, are believed to be a New World domestication, yet they have allegedly been found in ancient Chinese tombs! Similarly, the sweet potato, unquestionably native to the New World, was a staple in Polynesia before contact with the European explorers; in New Zealand it's known by its Maori name, *kumara*. The potato, native to South America, is sometimes incorrectly called the Irish potato but is nevertheless linked with Ireland because of that country's dependence on the potato and because of the famine in the 1840s when the potato crop failed.

One mystery, however, can be explained easily. A staple of the American holiday diet is the sweet potato, first domesticated somewhere between the Yucatán peninsula in Mexico and northern South America. Some, however, insist on calling these American plants "yams." Yams are of African origin and slightly resemble the sweet potato but are of an entirely different species. They're not available in most American supermarkets. Why the confusion? When new, softer varieties of sweet potatoes were first cultivated in the American south (especially Louisiana) growers attempted to differentiate these newer varieties from the older types by calling the new ones "yams." The US Department of Agriculture, which understands the difference between yams and sweet potatoes even if the rest of us don't, insists that anything called "yams" that are really sweet potatoes must be labeled as "yam sweet potatoes." **Answer 23b:** The geographic distinction is this: sweet potatoes originated in the New World; true yams originated in Africa.

The domestication of animals is also central to the Agricultural Revolution. Improved animal species not only gave mankind a more reliable food supply but also served as a dependable source of clothing and transportation. Domesticated animals are also associated with the Southwest Asian and New World hearths. **Answer 23c:** Cattle, sheep, goats, horses, and camels come from the Old World, while very few animal domestications occurred in the New World: the only significant examples are the llama and its cousin, the alpaca. (If you mentioned the Muscovy duck and the guinea pig as well, take extra credit!)

Taking into account the advantages of animal domestication and the availability of species, the lack of animal domestications in the Americas is surprising. In general, Old World peoples tended to domesticate a wide range of animals, then move the herds seasonally to obtain better pasture or more ample water supplies. In the New World, some animals moved in a similar fashion, but their movement wasn't controlled by man. Instead, Native Americans would follow these movements and in some cases center their societies on the hunting of a particular herd animal. The bison hunters of the Great Plains are a well-known example. Interestingly, the bison is now kept as a herd animal by ranchers, implying that it probably could have been domesticated by the same tribes that preferred to hunt it.

One of the more interesting New World–Old World comparisons in domestication involves the reindeer, or caribou. In the Old World Arctic, these animals were domesticated and herded by Arctic peoples stretching from Norway through Siberia. In Alaska, however, the animal was hunted and not domesticated. By the end of the nineteenth century, caribou herds and other supplies of game were inadequate to feed the native western Alaskan population. Officials introduced the domesticated reindeer to the area and brought in Scandinavian herders to teach reindeer herding. By 1930, the domesticated reindeer population had grown to over 600,000. Although reindeer herding in Alaska has diminished in importance since then, this remains the only example of nomadic herding present in the New World.

We don't often think of animals as weapon systems, but some animals qualify. The Arabian horse, noted for its speed and especially

its endurance, enabled the Arab/Islamic army originally led by Mohammed himself to attain the rapid conquest of North Africa and
most of the Iberian Peninsula beginning in the eighth century. One
legend has it that Mohammed allowed a herd of very thirsty Arabian
mares to race toward an oasis and then called them back. Five mares
returned at his call, and these became the breeding stock for his army's mounts. Interestingly, the Arabs used mares as their war horses.

One of the most feared animal weapon systems is the elephant.
Unlike Arabian horses, war elephants were always male. They were
probably first used in India, but their use spread widely. Alexander
the Great captured war elephants from the Persians and introduced
them to Mediterranean civilizations. The great Carthaginian general
Hannibal crossed the Alps with elephants in his failed attempt to
conquer Rome. The story is well known, but it's unknown where the
elephants came from. One elephant species was native to the Middle
East and North Africa, but that species (which may have been used)
is now extinct. Whether elephants from any part of the world have
actually been domesticated is debatable. Domestication implies selective breeding, and this brings about changes in the characteristics
of the animal. The Asian, or Indian, elephant has been tamed but
not really domesticated. Attempts to tame the African elephant have
been far less successful.

Domesticated animals that have been released back into the wild
are called feral animals and have been a continuing subject of heated
debate among environmentalists, stock breeders, and governments.
Perhaps the oldest example of feral animals is Australia's dingo, a
feral species of dog that has existed for around three thousand years
or longer. At various times, attempts have been made to eradicate
dingoes, to build fences to contain them, to prevent interbreeding
with other dogs, and to preserve pure genetic stock.

Spanish explorers, anticipating shipwrecks, intentionally released
cattle in various areas of the Caribbean and mainland America. The
idea was that shipwrecked sailors would find a ready supply of beef
were they lucky enough to wash up on shore. Incredibly, the idea

worked very well. Herds of feral cattle thrived in areas of the New World. Texas longhorn cattle are the direct descendants of feral Spanish cattle.

Feral horses that transformed the societies of several Amerindian tribes in the Great Plains and farther west were also gifts from the Spanish. Although some of these tribes had long specialized in hunting the bison, after 1750 their populations began to grow significantly, almost certainly as a result of greater food security from the use of the horse in pursuit of bison.

Camels are native to the New World but were never domesticated there. There is evidence that camels were still present in North America with the arrival of man and in fact may have been hunted to extinction. In a major irony, domesticated camels were introduced to the southwestern United States just before the Civil War; those responsible apparently had no idea they were returning the species to its ancestral home. Not surprisingly, camels did very well, but although they proved their usefulness they were abandoned and became feral. Camel sightings in the Southwest continued well into the twentieth century. Some still claim that a camel population survives in remote areas, particularly in Baja California, but there has been no authenticated sighting in over fifty years.

Interesting contemporary examples of feral animals include the colonies of feral cats that survive on and near certain college campuses. The cats presumably originated as pets of dormitory residents and are sustained by food provided by students and faculty. As with Australian dingoes, arguments rage about whether the cats are beneficial or harmful to the environment.

CONTINENTAL TIDBITS

Question 24a: *Which continent contains 70 percent of the world's freshwater?*

Question 24b: *What two countries occupy territory on two continents?*

Question 24c: *What is the second most populous country in South America?*

Question 24d: *What two countries in South America lack seacoasts?*

Question 24e: *What country was the* Titanic *closest to when she sank?*

Question 24f: *What self-governing religious enclave, part of a much larger country, fell under the special protection of Adolf Hitler during World War II?*

Question 24g: *Which is the only continent without active volcanoes?*

When geography is taught at all in elementary schools, one of the first things the students learn is the seven continents. It seems a reasonable starting point, but the whole idea of continents is a matter of custom and convention rather than of science or even of reasonable definition. We teach today that the smallest continent is Australia, yet geography texts of the 1920s taught that there were only six continents and Australia was the world's largest island. Why should Asia and Europe be considered separate continents when nothing really separates them? For that matter, why aren't Asia, Africa, and Europe (sometimes called the world island) all one continent?

However we may define continents, the fact remains that they are fundamental areal units of geographic study. One of the traditional definitions of the science of geography is "the study of the earth as the home of mankind." While of course some people do not live on continents, overwhelmingly mankind is confined to six of the seven continents. Along with oceans and seas, continents make up the big chunks in our mind's image of the earth. As we earlier learned in the case of Canada being south of Detroit, "big chunk" thinking can lead us astray. When we think of Asia, we visualize China and Japan, or maybe India. A country like Lebanon doesn't readily spring to mind, yet it's also in Asia. Similarly, Africa conveys the idea of herds of gnu and prides of lions, not necessarily the traffic congestion of downtown Cairo.

ASIA: WHERE MOST PEOPLE LIVE

A recent visitor from another galaxy would decide that the typical earthling is an Asian male. This is because a substantial majority of the world's population is found on the Asian continent, and, although females are a majority in many countries, they are a minority in Asia. Some of the highest national population growth rates are found in Asia as well. Presumably our visitor, however, would not be burdened with the same myths about Asian population that we carry around. Dozens of my students and colleagues have assured me that if the Chinese lined up and marched four abreast into the ocean, the line would never end. Even making some pretty generous assumptions, I could never get this to work out any way other than with the extinction of the Chinese population. The real surprise, however, is that currently the Chinese population is reproducing at a rate that, if continued, would lead to the disappearance of the Chinese without any need to have them march into the sea!

Chapter 20 indicated that the total fertility rate for China is only 1.54, well below the level needed (around 2.2) for the Chinese population to replace itself. If this level is correct and if it persists, we might expect the population of China to begin shrinking as soon as the year 2020! The major reputable sources of population projections all show China's population to begin to shrink no later than sometime between 2030 and 2040. This is so entirely contrary to conventional thinking about China that it may be the biggest surprise contained in this book!

Population geographers describe the major clusters of world population—areas of high population density extending over significant areas—with reference to the continents. The two largest are both in Asia. The larger of the two, the East Asian cluster (including much of China, as well as Japan and Korea), along with the South Asian cluster (much of India and Pakistan and all of Bangladesh), contain more than half of all the people on earth. While the growth of the East Asian cluster has slowed substantially in recent years, the continued rapid growth of the South Asian cluster will ensure that Asia continues to hold a majority of the world's people for the foreseeable future.

ANTARCTICA: WHERE NOBODY LIVES

In sharp contrast to Asia, which is the largest continent in both area and population, the fifth largest continent has no population at all. At any given time, Antarctica may contain about five thousand people living in several research stations, but in no way can they be considered a true "population." People can survive there, just as they can live in an orbiting space station, but they can hardly claim the Antarctic as their home.

Antarctica is increasingly becoming a tourist destination and is regularly visited by cruise ships. One of the earliest attempts at tourism, however, came to a tragic end in November 1979 when Air New Zealand flight 901 crashed on Mount Erebus, the most active of the Antarctic's volcanoes. This was a scheduled passenger flight that flew round trip from New Zealand to Mount Erebus for the sole purpose of sight-seeing. The flights were cancelled after the accident and have never resumed.

Answer 24a: The Antarctic continent has the highest mean elevation of the continents and, within its ice cap, contains 70 percent of the world's freshwater. It's often claimed that global warming could release this water, thereby causing dramatic rises in sea level. Antarctica is a true desert; it's simply too cold for much precipitation to occur. With warming, more precipitation might occur and the ice cap could initially grow rather than shrink. In turn, that would imply a drop in sea level, not a rise! We have lots of evidence of the warming of the planet, but the implications are less certain.

EURASIA: WHERE COUNTRIES STRADDLE CONTINENTS

Because our image of continents includes the idea that they are discrete units, we're surprised to hear that modern countries could straddle continents. Europe and Asia are in fact a single land mass, sometimes referred to as "Eurasia." Movements of European people to the east into Asia and the movement of Asian people westward into Europe has been going on for thousands of years. The arbitrary boundary between the two continents is in part the Ural Mountains. Despite this boundary line, Uralic people are the primary inhabitants of the European countries of Finland and Estonia, while European Slavs have migrated eastward to the Pacific shores of Asia. **Answer 24b:** Russia, predominantly a country of European Slavs, includes territory mostly in Asia but in Europe as well, while Turkey, predominantly a country of Asiatic Turks, is mostly in Asia but contains land in Europe as well.

Except for a nasty bit of political intrigue engineered by US president Theodore Roosevelt, there would be a third country that straddled two continents: Colombia. When the United States was contemplating completing a canal across the Isthmus of Panama, Roosevelt became impatient with negotiations with Colombia, within whose territory the isthmus lay. An independence movement erupted in Panama, supported by and largely staged by the United States. Panama declared its independence, and the United States

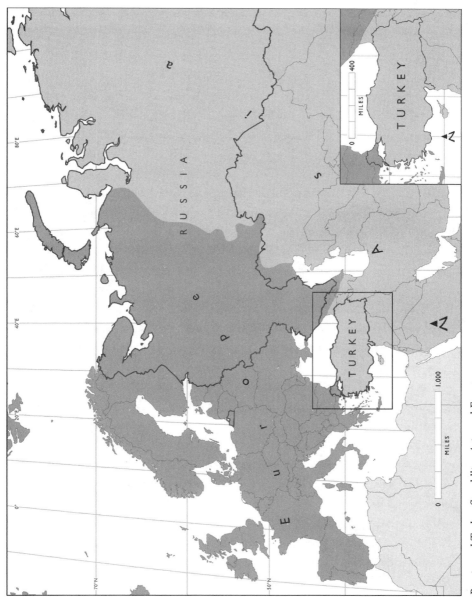

Russia and Turkey Straddling Asia and Europe

quickly negotiated a canal treaty with the new country. Panama is entirely in North America, while Colombia is now entirely in South America.

AFRICA: THE HOME OF MANKIND

Africa was commonly called the "dark continent" in the nineteenth century because so little was known about it. The term was used by a commentator on NPR (National Public Radio) in 2008 and drew a firestorm in response. Listeners apparently believed that it was a racist term, or otherwise demeaning of Africa. NPR apologized for its alleged "Africa bashing" (discussed in an earlier chapter). Unfortunately, Americans tend to remain in the dark about Africa. This is understandable because the United States has had less diplomatic, colonial, or military involvement with Africa than with any other region of the world. Still, more than 12 percent of the US population traces its ancestry to Africa. While growing Afro-American, or Black Studies, programs, in both secondary schools and colleges, have exposed students to neglected aspects of Afro-American accomplishments, contemporary Africa still seems overlooked.

About twelve thousand years ago, the last retreat of the continental glaciers was accompanied by the creation of the Sahara, the world's largest mid-latitude desert. Although Africa is the second largest continent, its southern two-thirds, particularly the interior regions, were isolated by the Sahara from the Mediterranean and European civilizations that developed to the north. This, of course, contributed to the "dark continent" idea and, in particular, meant that Europeans were ignorant of the cultures that developed in Africa south of the Sahara.

The evidence is convincing that Africa is where man originated. The famous fossil "Lucy" was a partial skeletal remains of an ancestor of man dated at over 3 million years old. More recently, two other hominid fossils have been found that date to the incredible age of 10 million years. All these fossils have been found in a geologic feature called the **EAST AFRICAN RIFT ZONE**. This region is a split in the African continental plate that has produced the highest mountain ranges in Africa and some of the deepest lakes in the world. The geology of the area has produced sediments that have preserved these ancient fossil remains.

> Dr. G says: Many publications refer to the "great African rift zone" or to simply the "Great Rift." This huge rift extends from Syria to Mozambique. The East African rift is considerably smaller.

SOUTH AMERICA: MANY COASTAL CITIES, TWO LANDLOCKED COUNTRIES

In some regions of the world, one country may dominate in terms of both population and area. Nigeria in West Africa and Indonesia in Southeast Asia are examples, but neither dominates to the same extent as Brazil does on the continent of South America. Brazil occupies almost half of the total area of the continent and just about half the total population as well. **Answer 24c:** With more than 190,000,000 inhabitants, it has more than four times the population of the second largest South American country, Colombia. Its largest city, São Paulo, is the largest city in South America.

All of South America was colonized by European countries at one time. Some of the continent still is: the French space program is centered in French Guiana! Because the European powers viewed the continent as a source of agricultural products and mineral resources, seaports were needed to facilitate trade. The population of South America is coastally oriented, with the interior of the continent considerably less developed. This factor led to Brazil constructing a new capital city, Brasilia, in the interior. **Answer 24d:** Despite the general coastal orientation of the entire continent, two South American countries lack seacoasts: Paraguay and Bolivia.

The Landlocked Countries of Bolivia and Paraguay

Bolivia's case is especially interesting because it once had a sea-coast but lost it in a war with Chile. Since the late nineteenth century, Bolivia has regularly sought to regain its coast. Chile, Peru, and Bolivia have squabbled over the towns of Arica and Talca, which have remained solidly in Chile for more than a century. In the 1990s and again in 2010, Peru offered to lease a small amount of land to Bolivia so that it could have a port on the Pacific. Perhaps this will work out, but Bolivia will need to make a substantial investment to make its leased seacoast into an actual seaport. Geopolitical writers like to point out that the issue of Bolivia's claim to a seacoast is one of the longest-standing border disputes in modern history.

Paraguay, aside from being landlocked, can lay claim to several unique distinctions. Its war against Argentina, Brazil, and Uruguay in the 1860s (the War of the Triple Alliance) may have been the bloodiest war in modern history. Exact population losses are uncertain, but Paraguay lost at least 50 percent of its entire population, and some estimates run as high as 80 percent. At the conclusion of the war, fewer than thirty thousand adult males were left alive in the country. Aside from the high mortality, there was only one adult male for about every five to ten adult females, a factor that obviously prevented normal family formation. If that were not bad enough, Paraguay literally lost its history in 1869 when an invading Brazilian army carried off all its historical records in the process of sacking and burning Asunción, the capital.

NORTH AMERICA: WHERE WAS THAT ICEBERG?

The North American continent is home to most of the readers of this book and so familiar to us that we hardly need to mention much about its geography . . . or do we? The point has already been made that the residents of the most populous North American country, the United States, ignore the largest country in area, Canada. Sometimes, however, there's a twist to this ignorance. At a trivia game aboard a ship on which I was lecturing, a question was posed about

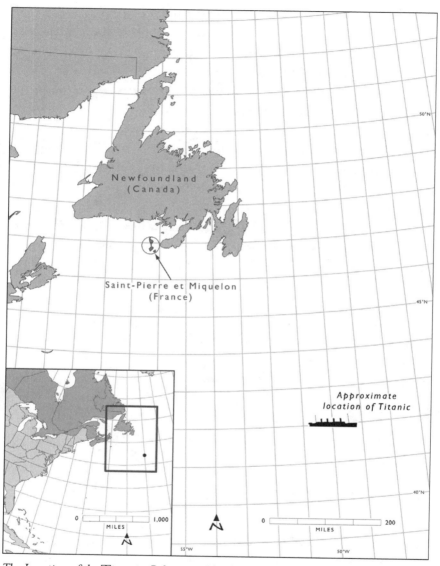

The Location of the Titanic *Relative to Newfoundland and St. Pierre and Miquelon*

the country closest to the *Titanic* when she sank. I cringed because I had heard the question asked before, and always the answer given by the game host or hostess was incorrect: Canada. This time, an elderly gentleman from Ontario (that's in Canada!) was outraged when the expected answer was given. Eventually he called the hostess an imbecile, and she called security. The audience was mystified. The site of the sinking has been mentioned in movies and books and on television. Besides, no country other than Canada was anywhere near the sinking, or so they thought. **Answer 24e:** The *Titanic* sank on April 15, 1912, "off the coast of Newfoundland," so the story line went. In 1912, Newfoundland was an independent dominion within the British Empire . . . and not a part of Canada. Newfoundland only became part of Canada in 1949, and even today, Newfoundland is not a Canadian province but merely part of one!

The situation could actually be worse! Just off the coast of Newfoundland are two small islands, St. Pierre and Miquelon, which constitute a department of metropolitan France. So, Newfoundland was the closest country to the sinking, but France may have been even closer than Canada! Remember the poor Canadian gentleman who may still be rotting in the ship's brig and remember to be genteel during trivia games, even when you have the correct answer!

EUROPE: THE BIG, THE SMALL, AND THE REALLY SMALL

Whether justified or not, Europe is "the continent" (as in, "We'll be touring the continent this summer"). It's the continent most visited by tourists from everywhere, and it may be the continent with the overall highest living standards (only Australia would compete). Its languages have been the world's international languages for the last two thousand years, and its inventions have profoundly affected the rest of the world. Its political geography, however, is a bit strange.

While Europe contains, in part, the largest country on earth, it also contains several of the smallest, most of which are answers to common geographic trivia questions. San Marino, Andorra, Monaco,

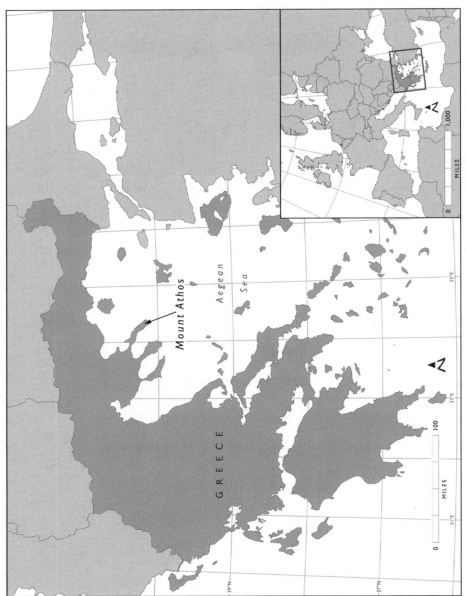

Mount Athos

Lichtenstein, Vatican City, and sometimes even substantially larger Luxembourg, seem anomalous in a world dominated by larger countries. **Answer 24f:** If you're looking for the self-governing religious enclave, part of a much larger country, that fell under the special protection of Adolf Hitler, do not look to Vatican City. Instead look to Mount Athos, a self-governing territory within Greece, where twenty Orthodox monasteries date their origins to the earliest days of Christianity.

AUSTRALIA: UNIQUE SPECIES

Australia is the smallest continent but certainly large enough to justify continental status. It's also the only continent occupied by a single country. Geographic isolation—aided by ocean currents and sailing conditions, the Great Barrier Reef, and the desert dominating most of the continent—has led to the evolution of unique and fascinating native species. The kangaroo, platypus, spiny anteater, and koala are well-known examples. So too are the most venomous snakes in the world, found both on land and in the surrounding oceans.

Answer 24g: All continents have volcanoes, and Australia certainly has its share. It does not, however, have any volcanoes that are currently active. The last volcano to erupt on the Australian continent did so about an estimated eight thousand years ago. Some oceanic islands owned by Australia do have active volcanoes, but there are none on the continent itself.

ISLANDS DIVIDED

Question 25a: What was the largest contiguous empire ever?

Question 25b: What was the second country in the Western Hemisphere, after the United States, to gain independence?

Question 25c: On what island was the earliest domestic cat found?

Question 25d: In what country do we find a minority group called the Burghers?

Question 25e: What two countries share the island of New Guinea?

The political organization of the earth's surface is a continuing focus of the science of geography. The different ways that man has organized space follows a historical continuum from simple to complex, yet even the oldest and simplest forms can still be found in existence today. In all cases, however, the ultimate goal is the same: to cordon off a portion of the earth's surface and govern it so that the inhabitants can survive and perhaps prosper. There are four basic systems for the political organization of space:

- Rudimentary organization: Most commonly found in extended families among people with relatively inefficient technologies, this system involves the need to use an undefined area of the earth's surface to provide the resources for survival. The land may be used for collecting or gathering or for relatively primitive systems of hunting and fishing. There are no formal boundaries of the area used; nevertheless, the area itself, however vague, and the resources will be defended, when possible, against intruders.

- Tribal organization: In this system, the territory used by the tribe has reasonably well-defined boundaries demarcated by prominent landscape features (such as rivers or mountain ranges). The members of the group may also be members of a family, but generally the number of people involved is larger than those utilizing rudimentary organization.

- Kingdoms or empires: The leaders of these entities—kings, queens, emperors, empresses—can be considered at least symbolically to own the land encompassed by the kingdom or empire. In the classic feudal (and similar) systems, the king would grant land and title rights in exchange for military, economic, and political support. Strangely, most modern kingdoms have limited the role of kings and queens to the status of figureheads, while some places that are not kingdoms operate much like old-style kingdoms used to but do not call their leader a king. Empires can be considered highly successful kingdoms that have extended their governance far beyond their original boundaries. The most powerful empires were the Roman and British empires, both of which extended to overseas areas. As the saying went, the sun never set on the British Empire so vast were its holdings, while Rome imposed its rule across the Mediterranean Sea and across the English Channel. **Answer 25a:** The largest contiguous empire, however, was the Mongol empire that extended from China to Europe and almost to Japan. The Mongols twice attempted to invade Japan but were thwarted by windstorms (probably typhoons) that the Japanese called "divine wind" . . . *kamikaze*!

- The nation-state: This system, in a sense, is a modern equivalent of tribal organization. It organizes space so that people with an affinity for each other (the "nation") are included in a bounded entity (the "state"). This seemed an improvement over kingdoms and empires, which traditionally were beset with a lack of uniformity in the form of many different nations within a single state. Most existing states are far from perfect nation-states, and some, as in the cases of Belgium and Canada, must constantly confront problems associated with governing more than one nation in a state.

At first blush it might seem easier to organize space politically on an island, and perhaps it is, at least when compared to a continent. After all, an island is smaller than a continent, and its boundaries are much more evident. Upon closer examination, however, on numerous islands it has proved very difficult to organize space politically in a way that brings the harmony and prosperity that are the goals of such organization. In fact, divided islands offer examples of some of the most contentious and bloodiest failures at political organization.

Great Britain offers one of the earlier examples of an island divided. The Roman Empire imposed its imperial rule there in the first century AD but was unable to extend it northward into a land it called Caledonia. In the second century, the empire built Hadrian's Wall across northern England. The obvious purpose of the wall was to keep the inhabitants of Caledonia, the Picts, whom the Romans were unable to conquer, north of the line and to prevent their raids south of the wall. Hadrian's Wall, however, contained gates and individual forts, so it's believed that another purpose was to impose taxes on trade coming from the north. Later, the Romans extended their control farther north into Caledonia (modern Scotland) and established a second fortified wall known as the **ANTONINE WALL**.

Dr. G says: The Antonine Wall was used only for a short period of time by the Romans, but its significance is that it marks the northernmost boundary of the Roman Empire.

The two walls approximate the division between England and Scotland, two areas of Great Britain that have fought so frequently and passionately with each other that at times it seems difficult to believe that they have been united in a single kingdom—the United Kingdom—since 1707. Even today, however, a visitor to Scotland is more likely to see a flag bearing the **CROSS OF ST. ANDREW** rather than the Union Jack flying over public buildings.

> Dr. G says: The flag of the United Kingdom, often called the Union Jack, is supposed to contain the Cross of St. George (representing England), the Cross of St. Andrew (representing Scotland), and the Cross of St. Patrick (representing Ireland). Whether there actually is a Cross of St. Patrick is open to question.

The United Kingdom added Ireland in 1800 but then broke apart in 1922 when the Irish Free State was created. For about two days, all of Ireland remained unified as a dominion within the British Empire (the same status, at the time, held by both Canada and Newfoundland). Then, Northern Ireland opted to separate from the Free State and remain within the United Kingdom. In 1937, the Free State became the sovereign state of Ireland. In 1948, Ireland became the Republic of Ireland and severed the few residual ties it still had with the United Kingdom.

Northern Ireland is often referred to as "Ulster," but the Irish province of Ulster contains nine counties, three of which are in the Republic of Ireland while the other six make up Northern Ireland.

It would be a gross oversimplification to claim that "Ireland divided" dates only to the creation of the Free State in 1922. Ireland's population composition and its political reality has been shaped and reshaped over hundreds of years. Nearly a quarter million people from what is now Northern Ireland migrated to the British colonies in North America before the American Revolution. In general, they were ancestors of "colonizers," mostly Presbyterians from Scotland, who had settled in Northern Ireland. In the colonial United States they were known as the Scotch-Irish, and they tended to avoid the British-dominated coastal areas and instead became the dominant population in Appalachia and the Ohio Valley.

Ireland is divided not only by the border between Northern Ireland and the Republic of Ireland but also by religious differences within Northern Ireland. Violence leading to more than three thousand deaths from the 1960s to the approximate present is referred to as "The Troubles" in Northern Ireland. The violence stems from religious differences between Protestants and Catholics who are also split between a desire to unify all Ireland within the framework of the Republic of Ireland (generally, Catholics) and those who favor a continued affiliation with the United Kingdom (called "Unionists," and generally Protestant). Terrorism and extreme political rhetoric characterized most of the last fifty years in Northern Ireland, but more recently grassroots efforts on both sides offer promise of a more peaceful future.

Hispaniola is a Caribbean island divided into the two countries of Haiti and the Dominican Republic. The first European colony in the New World was established on Hispaniola during the first voyage of Columbus in 1492 and became the base from which Spain built an empire in the Western Hemisphere. It was a source of enormous wealth not only for Spain but later for France as well. It was the first place in the New World where the Spanish found gold, but the real wealth was in sugarcane and indigo, which flourished there.

Disease decimated the native population of Hispaniola, and Spain began the importation of African slaves. A census taken near the end of the sixteenth century showed that slaves outnumbered the Spanish population by about ten to one. As Spain's empire expanded, its interest in Hispaniola diminished. In the early seventeenth century, Spain ceded the western part of the island to France. Spain's holding became known as Santo Domingo, while France's was called Saint-Domingue.

Saint-Domingue became the richest holding in the Caribbean and furnished much of the wealth with which the French king Louis XIV built the fabulous palace in Versailles, outside Paris. The wealth was based on one of the most extensive and brutal slave systems ever employed. Because so many slaves died from disease and harsh conditions, France was forced to bring in a continuous flow of new slaves from West Africa.

In 1791, a slave rebellion broke out in Saint-Domingue, inspired in part by the ongoing French revolution and aided by the French preoccupation with their internal problems. Later, France, under Napoleon attempted to regain control and reimpose slavery. Napoleon's army was not only defeated in battle but virtually wiped out by disease, especially yellow fever.

Answer 25b: Haiti became the second country in the Western Hemisphere, after the United States, to gain its independence. Simón Bolivar, the man who would become the liberator of much of South America from Spanish rule, took refuge in Haiti shortly after its independence and later received military support from the Haitian government.

Haiti has the distinction of being the only country to have been created from a slave revolt. It also has the dubious distinction of having declined from being probably the richest area of the Caribbean to becoming the poorest country in the Western Hemisphere. Its high population density today, its low per capita income, and recently, the earthquake that devastated the capital, Port-au-Prince, are viewed by its island-sharing neighbor, the Dominican Republic, as stimuli to border crossings and therefore as threats. The two countries have a history of bloody border conflict. Hispaniola is truly an island divided.

Cyprus, the third largest island in the Mediterranean, came under British control after the Ottoman Empire ceded administrative rights in 1878. Cyprus gained its independence in 1960, but soon afterward violence escalated between different national groups, Greeks and Turks, on the island. In 1974, Turkey invaded Cyprus and seized the northern portion of the islands. While the Republic of Cyprus claims sovereignty over the entire island, Turkey has established a de facto regime on the part of Cyprus it controls. An estimated 150,000 Turks have migrated from Turkey and settled in the Turkish-held region of north Cyprus, an action that not only reinforces Turkey's presence on the island but makes it likely that Cyprus will remain a divided island.

Conflict in Eastern Europe, Asia Minor, and Cyprus between Islamic Turks and Orthodox Christians is a continuation of a struggle that has gone on since the fifteenth century. The fall of Constantinople to the Turks in 1453 spelled the downfall of the Byzantine Empire and the beginning of ethnic hostility played out most recently in the "ethnic cleansing" that occurred in areas of Albania and parts of former Yugoslavia.

Some sources credit the Egyptians with first domesticating the cat, but DNA and archeological evidence now show that the cat was domesticated in the Middle East before civilization developed in Egypt. **Answer 25c:** In fact, the earliest evidence of the domesticated cat, dating to about **9 5 0 0   B C E**, has been found on Cyprus!

> Dr. G says: Dates are often said to be "before the common era" (BCE) rather than "before Christ" (BC).

Sri Lanka, known as Ceylon until 1972, is an island that has been divided in different ways and at different times in its history. Although a small population of indigenous people (called the Vedda) may still exist there, Sri Lanka was settled by the Sinhalese people from Northern India more than two thousand years ago, and they remain the majority population today. The principal geographic division is a climatic one: a dry zone in much of the northern half of the island and a wet zone in the south. An advanced civilization based on extensive irrigation systems developed in the dry one but mysteriously collapsed around 1,500 years ago. The dry zone became depopulated to the extent that some believe the population in Sri Lanka is smaller today than in 500 AD.

Sri Lanka has the unusual distinction of having been colonized by three different European powers: the Portuguese, the Dutch, and the British. The Portuguese and the Dutch never brought the entire island under their control. The central part of the island featured the Kingdom of Kandy, which resisted colonialism. This division of the island into colonized and uncolonized areas continued during the British era until the British conquered Kandy in the nineteenth century.

When Sri Lanka gained its independence in 1948, it was placed in an unusual situation by its colonial heritage. The Sinhalese, Buddhists for the most part, made up more than 70 percent of the population and, not surprisingly, gained political control of the newly independent country. The minority Ceylonese Tamils, who are predominantly Hindu, had been favored by the British and dominated the civil service. The best schools in the country also were concentrated in Tamil areas. Over time, various laws enacted by the Sinhalese majority deprived the Tamils first of their privileged position and more gradually of their ability to exist on an equal footing with the Sinhalese.

Eventually a Tamil rebellion broke out and resulted in the establishment of a de facto Tamil state in the northeast part of the island called Tamil Eelam. Civil war ensued. The conflict waxed and waned but continued for more than a quarter century. In 2010, the government of Sri Lanka declared the rebellion officially over, so in the strictest sense, Sri Lanka is no longer an island divided.

Several groups other than Sri Lankan or Ceylonese Tamils and Sinhalese are also well established in Sri Lanka. The Indian Tamils were recruited by the British to work on estates, or plantations, during the nineteenth century. Because the coffee estates were located in the hilly central area of the island, these Tamils are sometimes called hill Tamils. Disease ravaged the coffee bushes, and the area was replanted with tea, which today is Sri Lanka's most famous export.

After independence, Sri Lanka attempted to repatriate the Indian Tamils back to India, but eventually most were granted Sri Lankan citizenship.

Answer 25d: Descendants of Europeans from the colonial era also make up a distinct group within Sri Lanka's population. Although this number would include Portuguese, Dutch, and British, they're collectively referred to by a Dutch word: Burghers.

Three islands that are part of Indonesia are divided: Borneo, New Guinea, and Timor. The island of Borneo (or Kalimantan, as it is known locally) contains three political entities: the largest portion is part of the country of Indonesia; another portion, East Malaysia, is part of the country of Malaysia; and the third section is the independent country of Brunei. Borneo is the only island in the world divided into three separate states. **Answer 25e**: Indonesia also shares the island of New Guinea with the independent country of Papua New Guinea. At various times the Indonesian portion of New Guinea, the western half of the island, has been known as West Irian, Irian Jaya, and West Papua. Timor is divided between Indonesia and the now-independent country of East Timor; within the last decade it has been one of the world's major trouble spots.

These divided islands are legacies of Indonesia's colonial era, and at various times conflict has broken out between and within the constituent parts. In the 1960s, Malaysia and Indonesia were at war over the matter of East Malaysia. Although this matter has been settled by treaty, with the East Malaysian states of Sabah and Sarawak now firmly established on Borneo, there's an independence movement on New Guinea that seeks separation of western New Guinea from Indonesian rule.

East Timor was formerly a Portuguese colony and, along with the Philippines, one of the two areas in Southeast Asia where there was a predominantly **ROMAN CATHOLIC** population. In 1975, East Timor declared its independence from Portugal, but later that same year, Indonesian forces invaded and East Timor became an

Indonesian state. For more than two decades, a bloody, destructive battle for independence raged on the island. The United Nations intervened and a plebiscite was held in 1999. As a result, East Timor gained its independence again in 2002.

Dr. G says: Sources that cite East Timor and the Philippines as the only Roman Catholic countries in all of Asia are correct, but another Roman Catholic colony suffered the same fate that befell East Timor. Goa, another Portuguese colony, was absorbed by India, but unlike East Timor it neither sought nor attained independence.

Fiji is a divided island group that differs from other divided islands in the way the division occurs. Fiji received its first European settler somewhat later than other Pacific areas. It wasn't until the 1820s that the first Europeans came. In 1875, Fiji suffered from a measles epidemic that killed a third of its population. Only four years later, the British (to whom the islands had been ceded) began to import Indian laborers to work in the sugarcane fields. Indian-Fijians make up a little less than 40 percent of the total population, but they dominate in economic life and in urban areas.

Since Fiji became independent in 1970 it has faced a number of constitutional crises. The major split is an ethnic one, between indigenous Fijians and Indian-Fijians. Complicating this basic problem is the changing demographic situation: indigenous Fijians were in a minority but now are the majority, in part because ethnic tensions have caused the outmigration of Indians. Indian-Fijians, moreover, while controlling production (particularly the plantations), usually don't own the land but must lease it on short-term contracts.

OCEANS AND SEAS

....

Question 26a: Where are the highest tides?

Question 26b: What famous polar explorer was the first to take a vessel through the Northwest Passage?

Question 26c: The largest and smallest ocean border what three countries?

Question 26d: Where is the Sea of Tranquility?

Question 26e: Where is the Sea of Marmara?

Question 26f: What is the largest sea?

Question 26g: Where are the Islands of Langerhans and the Greater Palatine Canal?

....

About two-thirds of the earth's surface is covered with water, the majority of it in the form of oceans and seas. When I first crossed an ocean (the Atlantic) on a sailing vessel in 1960, I was impressed by the small Portuguese fishing boats that we occasionally encountered. These craft and their crews were glorified in Rudyard Kipling's *Captains Courageous* and, for the purposes of meeting a definition of *geography* proposed earlier ("the study of the earth as the home of mankind"), they spent enough time on the water and interacting with it that we could consider the ocean to be their home. Add to these few the groups of sea nomads found in Southeast Asia and others scattered on the oceans and seas of the world, and we conclude that there were very few such ocean residents in the past, and even fewer today. Basically, this two-thirds of the earth is nearly devoid of humans claiming it as their home.

The *use* of the oceans by man, on the other hand, has intensified over time. The Atlantic has served as the highway for one of the greatest human migrations in history: the movement of Europeans and Africans to destinations in the New World. We have no accurate count of how many undertook this voyage, but from the records we do have, the number must exceed a hundred million. In earliest times the ocean was crossed in vessels that seem much too small for the journey. Viking longboats and Columbus's *Niña* look puny indeed when compared with the US aircraft carrier *Ronald Reagan* or the cruise ship *Oasis of the Seas*, both among the largest vessels afloat today.

TIDES: LUNAR LIFTS

Some argue that we know less about our oceans and seas than we do about outer space, but we've nevertheless learned a good deal about them over time. We've long known about tides: what causes them, how to predict them, and how to use them to aid navigation. The first off-campus college class I taught was to submariners at Pearl Harbor. These students had a genuine interest in tides and I was surprised that they knew so little about them. Tides are influenced by the gravitational pull of the moon and sun and by the rotation of the earth. Aside from these basics, I also taught them about the US seaborne invasion of Betio and Tarawa Atoll during World War II, an incident in which the US Navy misjudged tides and nearly botched the landing. Landing craft were left stranded on reefs and sandbars when the calculation of tides proved to be tragically wrong. Only the extraordinary courage of the US Marines and some **UN-EXPECTED GOOD LUCK** led to a successful operation.

Dr. G says: Because the initial assault waves on Tarawa held such a precarious position after the first day of fighting, a Japanese nighttime counteroffensive might have been decisive. A fortunate American shell, however, killed the Japanese commander, and the entire Japanese command structure was compromised, so a counterattack never occurred.

Answer 26a: It seems common knowledge that the tides in the Bay of Fundy are the highest in the world. As with many other examples of so-called common knowledge, this may be only partially correct. The Bay of Fundy is the crease of ocean between eastern New Brunswick and western Nova Scotia, Canada, and including a portion of coastal Maine. Tides here reach more than fifty feet between high and low water. These tides have been known for centuries and have become quite famous. More recently, however, the tides at Ungava Bay in northern Quebec, Canada, have been measured at approximately the same levels as those in the Bay of Fundy.

OCEANS: LARGE AND SMALL

The largest ocean is, of course, the Pacific. From the European perspective, the Spanish explorer Balboa discovered it when he crossed the Isthmus of Panama on foot. He named it the "Southern Sea" because it lay to the south of the isthmus. (The peculiar shape of Panama has led to one of the most confusing aspects of the Panama Canal. To enter the Pacific through the canal, a ship must go south and slightly east from the Atlantic or **CARIBBEAN** side. Conversely, a ship travels slightly westward to enter the Atlantic.) Magellan later named the southern ocean *pacifico* ("peaceful") because, according to some accounts, he encountered favorable winds there. The Pacific is the largest geographic feature on the planet and is larger than all land features combined.

Dr. G says: The history of the Panama Canal says it connects the Atlantic and the Pacific. Maps say it connects the Pacific and the Caribbean Sea. The distinction is arbitrary.

The smallest ocean is the Arctic. It's also the shallowest and least salty. Only recently in world history have we learned much about the Arctic Ocean, including its existence. During the age of exploration, European cartographers were uncertain whether the north polar region was land or water. Over time, more and more maps showed it as water, not so much because there was evidence to show this, but because European sea powers longed for the existence of a short water passage to China. In fact, about 50 percent of the Arctic Ocean is frozen, even in summer, so the presence of an ocean didn't necessarily mean a feasible route to China.

By the end of the nineteenth century there were persistent rumors, and even eyewitness reports, that there was open water at the North Pole. Unbelievably, an 1893 geography textbook described the presence of open water. The Arctic Ocean was thus conceived as an ice doughnut, with open water in the "hole." **Answer 26b:** There was no open water, no shortcut to Asia from Europe via the Arctic Ocean . . . until the famous explorer Roald Amundsen (first to reach the South Pole) found it between 1903 and 1906. Global warming has reduced Arctic ice in recent years, and this Northwest Passage has been open sporadically to ship passage. In the summer of 1958, the USS *Nautilus* accomplished the task in record time by going from the Pacific to the Atlantic under the ice of the Arctic Ocean in less than one hundred hours.

Answer 26c: Only three countries in the world share boundaries with both the largest (Pacific) and the smallest (Arctic) ocean: the United States, Canada, and Russia.

SEAS: WHAT ARE THEY?

Arab folklore and the tales of Sinbad the Sailor mention seven seas, possibly describing the sailing route from the Persian Gulf to China. "Sailing the Seven Seas" seems to be a mythical term implying widespread oceanic voyaging destined to the equally mythical four corners of the earth. There are, however, more than one hundred geographic features in the world described as seas. Perhaps the most common use of the term *sea* is to describe a specific lobe or region of an ocean. The Baltic, Mediterranean, Caribbean, and North Sea are thus seen as divisions of the Atlantic Ocean. Other "seas," however, are not connected to oceans and might otherwise be described as lakes. The Dead Sea, the Caspian Sea, and the **ARAL SEA** (if it still exists) are examples. The Sea of Galilee, familiar to Christians, is only a small lake and could qualify as the smallest body of water to be called a sea.

> Dr. G says: The Aral Sea, once one of the largest freshwater lakes in the world, lies between Kazakhstan and Uzbekistan. The Russians diverted feeder rivers so that the lake is now only 10 percent of its original size. It's been called one of the world's worst environmental disasters.

Map users sometimes encounter places that seem to be both seas and non-seas! The Sea of Cortes, for example, is also commonly called the Gulf of California. The Great Lakes taken collectively, or considering only Lake Superior, are sometimes considered to be inland seas. They're certainly bigger than other bodies of water called seas. One of the most famous seas is a non-sea, or at least it doesn't contain water. **Answer 26d**: The Sea of Tranquility is on the moon, where craters were labeled "seas" by early astronomers. The **SEA OF TRANQUILITY** is also the site of the first lunar landing.

Dr. G says: The study of the moon as the home of man should properly be called *selenography*, although I've never heard the term used! Three prominent selenographic features of the Sea of Tranquility are named Armstrong, Collins, and Aldrin after its first explorers.

A sea is part of the dividing line between Europe and Asia. **Answer 26e**: Although the Ural Mountains separate Asiatic Russia from European Russia, the Sea of Marmara separates Asiatic Turkey from European Turkey. This sea connects two other seas: the Black Sea and the Aegean Sea. Strangely, the straits at either end of the Sea of Marmara, the Bosphorus and the Dardanelles, are better known than the sea itself.

The boundaries separating sea from sea or sea from ocean are often indistinct and arbitrary. For this reason, determining the largest sea involves assumptions about the actual boundaries. **Answer 26f**: The South China Sea is sometimes cited as the largest, but my estimate is that either the Coral Sea or the Philippine Sea are the biggest. One geographic error that the US State Department has been guilty of for decades is easy to avoid: migrants coming to the United States from Cuba or Haiti are not "trans-Caribbean migrants." The body of water separating the Greater Antilles from the United States is the Atlantic Ocean. The Caribbean Sea is south of Cuba and Haiti.

Geography departments frequently receive telephone calls and emails requesting geographic information. Almost all these questions can be answered by recourse to a map, but a few are a bit more obscure. **Answer 26g**: Several times I've been asked about the Islands (or Islets) of Langerhans and the Greater Palatine Canal. Those asking had the wrong university department. They should have been calling the biology department, as these are parts of the human body!

BRIDGES

Question 27a: The "Bridge of No Return" joins what two countries?

Question 27b: What bridge was called "Galloping Gertie"?

Question 27c: What bridge is noted for "caisson disease"?

Question 27d: Where is the longest bridge in the world?

An important topic in geographic study is connectivity: how interaction between places occurs and how this interaction is affected by changes in transportation systems. It's useful to consider the history of the United States and the political philosophies that shaped the development of the country from the standpoint of connectivity. Jefferson's vision for the country was for a largely rural population, secure on its own agricultural land, with minimum connectivity needed only for the export of agricultural crops. This situation, a largely rural setting with minimal connectivity, of course, was an essential reality in Virginia and the South. In the North, where Alexander Hamilton's thinking mirrored conditions there, maximum connectivity was needed to encourage commerce, trade, and, eventually, large-scale manufacturing. Perhaps it's mere coincidence, but famous bridges, icons of connectivity, were concentrated in the North.

The one bridge I will never forget is the mythical bridge used by the Three Billy Goats Gruff. As I explained to a summer class of elementary school teachers, the Billy Goats Gruff story is really an allegory about a particular type of connectivity, the practice of "transhumance," the movement of herd animals seasonally to better pasture. In alpine Europe, transhumance is vertical, with the animals going up and down slope with the seasons. The troll in the story represents a border guard, impeding connectivity. Unfortunately, national borders in several parts of the world interfere with transhumance, and like it or not, herd animals tend to ignore borders, just as the story shows. One teacher took exception to my choice of story and complained to university officials that I was teaching "something that advocated violence as a solution to disagreements." While I did offer to forward her complaint to the Biggest Billy Goat Gruff, I found this somehow only made the situation worse.

Bridges are often associated with wars. From Horatius, to the bridge at Owl Creek, to the bridge at Remagen that the Germans forgot to blow up, to the one on the River Kwai, and to the bridge that was too far at Arnhem, war and bridges go together because of connectivity. An army on the advance wants maximum connectivity, while one retreating wants no connectivity (behind its line of retreat). One of the most memorable "war bridges" involved the repatriation of prisoners of war in 1953 after the cessation of hostilities in the Korean War. **Answer 27a:** The "Bridge of No Return" was built through a demilitarized zone connecting North and South Korea. Prisoners could choose to cross the bridge or stay where they were, but once they crossed they couldn't return. The bridge was used again when the crew of the US Navy ship *Pueblo* was repatriated after capture by the North Koreans. It was an example of one-way connectivity.

In their heyday, the Romans were the greatest military power to be found anywhere. Since military power depends on good connectivity, it's not surprising that the Romans were the greatest bridge and road builders until modern times. They used the principle of the arch to build their bridges, and they were the first to use concrete. I personally find it disconcerting that the road I customarily took to my university was always in various stages of disrepair, while some Roman roads are still in use after two thousand years! Even worse, some Roman bridges are still used, while some of ours have already collapsed.

Answer 27b: Perhaps the most famous failed bridge in American history was the bridge across the Tacoma Narrows, Puget Sound, in Washington State. This bridge, with the infamous nickname "Galloping Gertie," collapsed less than a year after it was built. It was an impressive structure when it was completed in 1940, one of the longest suspension bridges in the United States. Even while it was being built, workmen noted its tendency to buckle in relatively light winds. Engineering students have been enthralled for decades with the peculiar stresses induced in the bridge that made it appear to "gallop." Fortunately, when the bridge did fall, there was no loss of human life, although one panicked cocker spaniel couldn't be rescued from a car trapped on the structure.

Many great bridges have been built in the United States both before and after the demise of Galloping Gertie. At the time it was built, however, nothing compared to the Brooklyn Bridge. It was by far the longest suspension bridge ever built at the time of its opening in 1886. It greatly increased connectivity between Manhattan and Brooklyn and even more distant areas of Long Island. Its strength and durability were such that it still receives heavy use more than 125 years after its opening.

The Brooklyn Bridge, however, wasn't an easy bridge to build. It was originally intended that its support towers be built on bedrock, and this required laborers to work in pneumatic caissons, compartments below the level of the East River, which were filled with compressed air to keep the water out. Workers leaving the compressed environment quickly were likely to have nitrogen gas bubbles develop in their bloodstreams. This condition, originally called "caisson disease," is more commonly called "the bends" or "decompression disease." **Answer 27c:** So many workers died or were injured by caisson disease during the building of the Brooklyn Bridge that it was decided finally not to build the supports on bedrock but instead on rubble about thirty feet above bedrock.

Europe has built some spectacularly beautiful bridges in recent years: one across the Seine near its mouth and another from Sweden to Denmark. Presumably the design of these bridges has been tested in wind tunnels to avoid another Galloping Gertie situation. Given the incredible economic strides that China has made in recent years, it's not surprising that China has greatly increased its connectivity by building bridges. Trivia experts like to argue about how the length of a bridge should be measured: should it be the length of the center span? Should approaches be included? Should it be the length from shoreline to shoreline? **Answer 27d:** No matter how a bridge is measured, the longest bridge is currently in China. The most likely candidate is the Danyang-Kunshan railway bridge, which carries high-speed rail traffic between Beijing and Shanghai.

A tunnel that carries traffic is the mirror image of a bridge, and it's usually more expensive to build than a bridge. Recent tunnels, however, are not only impressive engineering feats but have been built in areas where bridges are impractical. The longest traffic-carrying tunnel in the world is under construction, but when completed the Gotthard Base Tunnel will be about thirty-five miles long under the Alps in Switzerland. The longest tunnel currently in operation is only slightly shorter: the Seikan Tunnel carries rail traffic between the Japanese islands of Honshu and Hokkaido for about thirty-four miles. The famous "chunnel" connecting Britain and France underneath the English Channel is about thirty-one miles long.

RIVERS

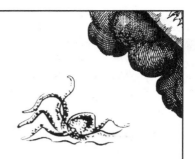

....

Question 28a: What river were Roman generals not allowed to cross with their armies?

Question 28b: What river was instrumental in General MacArthur's removal from command

Question 28c: What city is at the point where the Blue and White Nile rivers meet?

Question 28d: What is the longest river in Europe?

....

The earliest civilizations were built around rivers: the Nile, the Yellow River, the Tigris–Euphrates, and the Indus–Ganges. The reasons are fairly obvious: abundant freshwater, rich alluvial soil, transportation, defense, and fishing resources. Almost as soon as population grew along these and other important world rivers, however, they were turned into open sewers. While some of the advantages of river civilizations continued, drinking water and fishing resources were diminished, if not destroyed, by pollution. We have rescued some rivers. The Seine, at least the portion that runs through Paris, became so polluted that its stench drove French royalty away from its banks. Now, however, fishermen and swimmers safely use the river in the heart of Paris. The Cuyahoga River, running through Cleveland to

Lake Erie, is probably the most famous polluted river in the United States. Its notoriety spread in **1969** when the river caught on fire! While the Cuyahoga is not entirely clean today, it's much improved.

Dr. G says: The Cuyahoga has actually had several fires over the years, but the fire in 1969 became iconic and is often mentioned as instrumental in American public perception of the need to battle water pollution.

Political geographers in the past proclaimed the virtues of rivers as natural boundaries between countries. Rivers, they said, were clearly visible lines across a landscape, so there could be no "boundary disputes." Moreover, they claimed, rivers provided a measure of defense. An attacking army could at least be stalled by the need to cross a river. History was on their side because rivers have long been boundaries between different groups of peoples. The Rhine River, for example, has separated Teutonic speakers from Romanic speakers, or more specifically the French from the Germans, and has been a focus of conflict for hundreds of years. In the United States, the Rio Grande (or Rio Bravo) separates a portion of Mexico from Texas, while a portion of the St. Lawrence River separates New York from Canada.

The "river as boundary" idea was probably never a good one. A boundary, like a line in a coloring book, is supposed to separate like from unlike. People who live on the banks of a river are necessarily going to have more in common with each other than with those on either side who live far inland. Jobs, food supply, transportation, and a whole way of life center on the river. Rivers, moreover, are not immutable boundaries. Mature rivers change course frequently, and when they happen to serve as boundaries, every change in course means a border dispute. Ironically, while Mexico and the United States have spent millions of dollars adjudicating the common boundary along the Rio Grande, millions of people have crossed illegally into the United States from Mexico. Today, in an age dominated by aerial warfare, rivers offer only minimal defense against a military invasion.

Two rivers, two famous military commanders, flushed with stories of victory. Both commanders ignored standing orders not to cross a river. The one who succeeded became one of the most powerful dictators in history. The other was removed from command.

Answer 28a: Julius Caesar, after a successful campaign in Gaul, took his victorious XIIIth Legion across the Rubicon River, the boundary line separating **CISALPINE GAUL** and Rome proper. According to Roman law, any commander who did this was guilty of a capital offense, and so were his troops if they continued in military service. Caesar won the ensuing civil war and maintained control of the vast Roman Empire until his assassination.

> Dr. G says: *Cisalpine* simply means "this side (the Roman side) of the Alps."

"Crossing the Rubicon" has entered our language as an expression of total commitment. Unfortunately, and perhaps indicative of the modern meaning of *commitment*, no one is exactly sure where the Rubicon is or was. Changing drainage patterns on the landscape and perhaps modern excavation and construction have eliminated the Rubicon . . . or perhaps it's an existing stream with another name.

General Douglas MacArthur's military career was nothing short of incredible. From his time at West Point, where he finished at the top of his class, through his leadership of the occupation of Japan after World War II, MacArthur could lay claim to being the greatest military leader in American history. The Korean War, however, proved his undoing. MacArthur was placed in command of US, UN, and Republic of Korea forces during the Korean War. Following the entrance of Chinese troops into North Korea and the retreat of MacArthur's forces, President Truman relieved him of command.

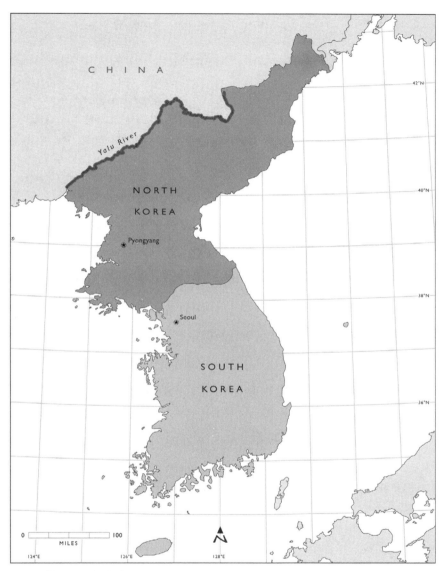

Korea Showing the Yalu River

Answer 28b: There is presumptive evidence that MacArthur goaded the Chinese into entering the war and that bombings north of the Yalu River, the border between China and North Korea, were contrary to orders given to MacArthur. Although MacArthur returned to the United States a hero, in a short period of time his image became tarnished as new evidence emerged. In particular, MacArthur was criticized for refusing to meet with President Truman on US soil and for commanding the war in Korea from a desk in Japan.

According to most sources, the Nile is the longest river in the world. After 1869, however, the Suez Canal was the most important waterway in Egypt. Although the French had built the canal, the British benefited the most. For the British the world shrank dramatically after 1869 because their most important colony, India, as well as Malaya, Australia, and New Zealand, became thousands of miles closer to Great Britain. The British Empire was already the largest political, economic, and military force in the world. The Suez Canal increased its power.

Egypt itself was not part of the British Empire when the Suez Canal was built. Egypt was then an autonomous holding of the Turkish Ottoman Empire. In 1882, the British landed an expeditionary force at both ends of the Suez Canal for the ostensible purpose of bringing political stability to the region. At that point, Egypt became a puppet state of the British Empire, although it remained de jure part of the Ottoman Empire.

Also in 1882 and, at least in part, as a response to growing British and European influence in Egypt, an Islamic leader arose south of Egypt, in the Sudan, and recruited thousands of followers for the purpose of ridding the area of Christian influence. The leader called himself "the Mahdi," a term reasonably close to the Judeo-Christian concept of a Messiah. The Mahdi is believed by Muslims to be one who will restore Islam to its former glory and destroy evil. Numerous individuals have arisen in history and have claimed to be the Mahdi.

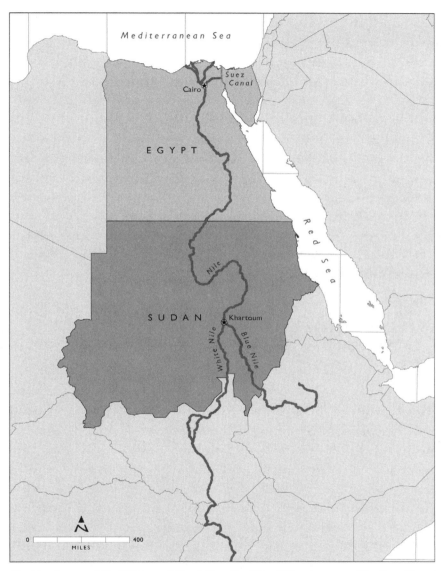

The Confluence of the White and Blue Niles, the City of Khartoum, and the Suez Canal

As the Mahdi's power grew, the British government decided that it would be in its best interest to evacuate its troops, allies, and civilians from the Sudan. **Answer 28c:** To accomplish this, in 1883 General Charles George Gordon was sent to the Sudanese capital, Khartoum, at the confluence of the White and Blue Nile rivers. Gordon, also known as "Chinese Gordon" for the victories he had won in China, was one of the most famous and popular military figures in Britain at the end of the nineteenth century. General Gordon succeeded in evacuating many Europeans down the Nile to Egypt, but gradually the Mahdi's armies surrounded Khartoum and captured it in 1885. Gordon was killed and beheaded. The British relief army arrived two days too late. This was an almost **UNBELIEVABLE DEFEAT** for the British, coming as it did during the height of their prestige and power.

Dr. G says: The nineteenth-century fall of Khartoum even today fuels radical Islam. It is surprising that prominent current leaders have not claimed to be the Mahdi.

Trivia questions about rivers can be frustrating. Through the use of satellite imagery and other techniques of remote sensing, it has been possible to better determine the length of rivers. Even so, there's no easy way to measure total length because the starting point is often unclear (e.g., Does the Mississippi actually start in Lake Itasca, Minnesota, or does it start somewhere in Montana or western Canada as part of the Missouri-Jefferson river system?) because rivers change course and because deltas are constantly changing as the river's "load," or soil and silt, is deposited.

Meanwhile, the so-called knowledge that generations of students have learned about rivers has become embedded in our minds . . . and it may be wrong. The amount of water flowing in rivers changes as well. There are seasonal changes, of course, but over time several rivers have seen major changes in their discharges. The Loire and the Tiber were once navigable but are no more, while the Colorado River, which once commonly flooded the Imperial Valley in California, loses almost all its water before reaching its outlet. A variation on this theme is that a wrong answer is given in a trivia contest and then passed on by word of mouth until it becomes embedded (but wrong) knowledge!

Strangely, one commonly asked river question seems to be always accompanied by the wrong answer even though there's not the slightest doubt about the correct answer. **Answer 28d:** The longest river in Europe is not the Danube but the Volga. The Volga also has the greatest drainage area and the greatest discharge volume. Since the Volga was a Soviet river, perhaps it was a victim of the cold war! If that's the case, it has certainly been rehabilitated, since its most famous city and the site of the turning point of World War II in Europe, Stalingrad, has been renamed Volgograd.

MEDICAL GEOGRAPHY

Question 29a: What disease was known as the "Plague of the Americas"?

Question 29b: What was the black death?

Question 29c: What disease decimated the population of Sri Lanka's dry zone?

Question 29d: What is the leading cause of death in southern Africa?

A medical doctor with whom I briefly worked told me that 1948 was the most important year in medical history. Before then, he told me, whenever a physician actively intervened with a patient's disease, the patient was likely to get sicker; after then, odds were that intervention improved the condition of the patient. We can look at this from a broader perspective and say that, for most of human history, we were disease conscious; now we're health conscious. Infectious diseases are not as omnipresent as they were in the past, yet they linger, and the threat exists that new strains of old diseases will return and overwhelm treatment options. (See chapter 44 concerning influenza.)

As world population grew and as connectivity between places improved, it became possible to spread contagious diseases over wide areas and to infect large numbers of people. In other words, the concept of epidemic became possible. The origin and spread of contagious diseases is a classic geographic problem, and it's not surprising that the geography of disease was studied long before we understood what caused epidemics or how to treat the diseases. The incidence of cholera, for example, was mapped more than a century before Pasteur advanced his germ theory of epidemic disease.

The age of exploration and colonization carried the Old World's epidemic diseases into the Americas and the Pacific, where such diseases were unknown and where native peoples had no natural immunity. Smallpox, tuberculosis, and even diseases that Europeans considered minor or to be "childhood" diseases became devastating to those who hadn't known them previously. A conservative estimate would be that at least half of Native Americans and Pacific Islanders were gone within a century after first contact with Europeans.

One of the worst killer diseases in the Western Hemisphere, ironically, killed Europeans and their descendents just as readily as it killed Native Americans. **Answer 29a:** Originating in West Africa, yellow fever was called the "Plague of the Americas." It was carried to the New World by the slave ships that transported mosquitoes, and it was found in the blood of slaves from that region, many of whom were at least partially immune to the disease.

Yellow fever was not just a tropical disease. New Orleans and Memphis both had devastating epidemics of yellow fever in the nineteenth century, although the United States began seriously to confront the disease only after American troops contracted it in Cuba during the Spanish-American War. Walter Reed proved that the mosquito was the vector of the disease. Armed with that knowledge, George Gorgas, my nominee as America's most unsung hero, went after the mosquito in Cuba and succeeded in eradicating both malaria and yellow fever. When the United States took over construction of the Panama Canal, Gorgas accomplished the same miracle in Panama despite widespread opposition from engineers, politicians, and doctors.

Yellow fever's dangers were not well understood by the medical community mainly because some outbreaks were mild and resulted in few deaths. Other outbreaks, however, had mortality rates of 50 percent or higher. Because mosquito-borne malaria had been discovered earlier to be a parasitic disease, some doctors believed yellow fever also to be parasitic. Others, however, considered it a contagious bacterial disease passed from human to human. Amazingly, some authorities still think yellow fever is parasitic, even though a Nobel Prize was given to Max Theiler for proving that it was a virus!

The black death is estimated to have killed half of Europe's population in the fourteenth century. Most authorities thought the disease was bubonic plague, but some argued that historic accounts of the disease didn't match the symptoms or the pattern of the contagious spread of plague. A mass grave of black death victims was recently disinterred, hundreds of years after burial. Evidence of the virus that causes bubonic plague was still present and was identified. **Answer 29b**: We can now say that evidence overwhelmingly shows that the black death was bubonic plague.

Fleas carried and spread the plague, but a variety of animals can carry the fleas. The ubiquitous rat population of Europe gets most of the blame. From the time of the early Middle Ages in Europe, the dominant rat was *Rattus rattus*, commonly called the black rat in Europe. This species unquestionably carried the fleas that carried the disease. The black rat, moreover, lived in close contact with the human population. Perhaps as early as the sixteenth century, a westward migration of a larger rat, *Rattus norvegicus*, began from northern China or Siberia. This rat, usually called the brown rat (even though most of them are gray) preens its coat and will do its best not to carry fleas. It replaced the black as the dominant rat population in Europe. Did this gradual replacement account for the decline in the incidence of bubonic plague in Europe? It may have, but we don't have enough evidence to be sure. Some have argued that the timing of rat movement doesn't coincide with the diminution of plague, but

while we know when plague broke out we have only sketchy information on the rats themselves. It would be a good study for a medical
geographer to undertake!

The island of Ceylon, now the country of Sri Lanka, was a legendary land in the mythology of both South Asia and the Middle East.
Supposedly, it was where the princes of Serendip lived, from which
we get the word *serendipity*. Even today, after decades of civil war, Sri
Lanka still conveys the image of a tropical paradise. The northern
part of the island is the dry zone, which contains extensive ruins
of irrigation systems and cisterns. This discovery has led some to
conclude that Sri Lanka's ancient population was much larger than
it is today. What might have caused the depopulation? **Answer 29d:**
Malaria is almost certainly the answer. In the 1930s and up until the
end of World War II, medical evidence indicated that malaria was
hyperendemic in the dry zone (which means that just about everyone
had it). Malaria is an ancient disease, but it's not native to Sri Lanka.
Its entrance onto the island seems to correspond to the beginning
of depopulation.

Malaria (unlike yellow fever) is a parasitic disease. At least 200
million cases occur each year, resulting in at least 700,000 deaths.
Of its four major forms, falciparum is the worst type of malaria. A
good parasite doesn't kill its host; in fact, its continued existence
depends on the survival of the host. Falciparum, however, is an
imperfect parasite that often kills its host, especially infants and
children. While malaria is certainly not unknown in Sri Lanka's
wet zone, the steeper topography and the greater amount of rainfall
there don't provide ideal breeding grounds for the malaria-carrying
mosquitoes (*Anopheles culicifacies*). The dry zone, however, particularly
with its man-made irrigation ditches and cisterns, allows for water
to pond—ideal breeding grounds for the mosquito.

In 1946, while Great Britain still ruled Sri Lanka, medical and
public health practices and supplies developed during the war were
made available to the colony. Three things in particular came to Sri

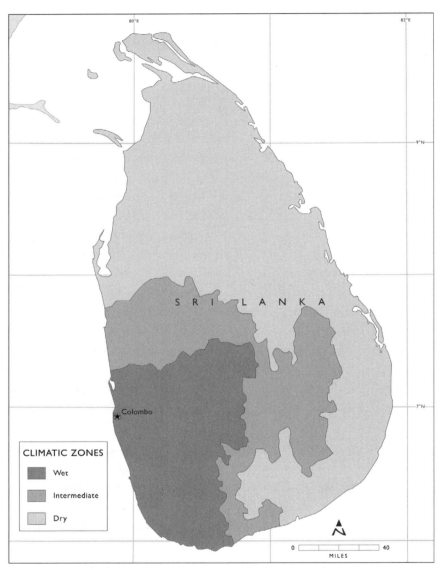

Wet and Dry Zones of Sri Lanka

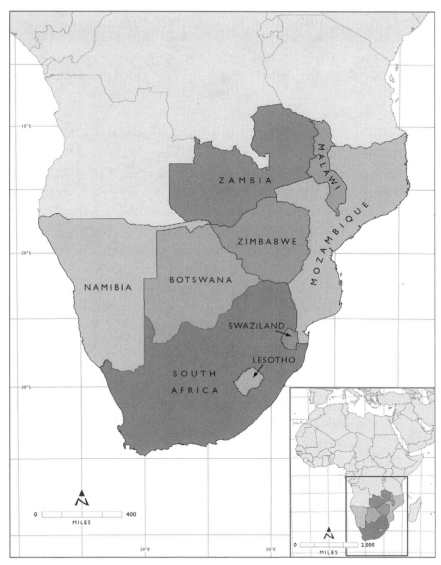

Southern Africa's National Boundaries

Lanka: the insecticide DDT, antimalarial medicines, and infant and child health clinics. Amazingly, in a single year, 1946, Sri Lanka's death rate was cut in half! Although at the time many were comforted with the thought that malaria had been eradicated from Sri Lanka (and India as well), this was not the case. Malaria continues to be a serious problem in both countries.

Malaria and infectious diseases are also common in tropical Africa. **Answer 29d:** However, it appears that recently HIV/AIDS has become the leading cause of death in the nine countries that make up southern Africa. Because HIV/AIDS is a disease passed on by sexual activity, it affects its victims in the prime of life, and the resulting mortality leaves behind dependent children and elderly. This region contains only about 2 percent of the world's population but almost a third of the world's HIV/AIDS cases. Projecting the HIV/AIDS epidemic in southern Africa is difficult. On the one hand, there's some evidence that disease rates may be slowing, but contrary opinions suggest that the disease is still on the increase and that continued high rates of the disease could lead to a general societal breakdown.

HIV/AIDS in southern Africa differs from the disease in other areas of the world because of the means of transmission. In most areas, the disease is associated with homosexual practices, with the reuse of needles and syringes by drug users, and with contaminated blood used in transfusions. In southern Africa, however, it's usually transmitted by heterosexual practices and between married couples. The blasé attitude of political leaders toward the disease has also interfered with attempts to slow the epidemic and provide treatment.

BORDER ISSUES

Question 30a: Along what river was the Siegfried Line built?

Question 30b: What is the longest unfortified boundary between countries?

Question 30c: With what border is the Zimmerman telegram associated?

Question 30d: What river, which empties into the Baltic Sea, is the eastern border of Germany?

Boundaries and borders are everywhere! A television science fiction series features battles over the **NEUTRAL ZONE** (presumably the borders are three-dimensional), a school district line may separate a **GOOD SCHOOL DISTRICT** from a poor one, and neighbors may quarrel over a property line as if their lives depended on the outcome. As a famous politician put it, "All politics are local," implying that local pride and a sense of belonging often come in small units. Affinity for local sports teams is a well-known example, but a community near my home features T-shirts with their zip code!

> Dr. G says: The neutral zone of science fiction fame was a real-world area in the desert lands between Saudi Arabia and Iraq.

Dr. G says: There have been two recent arrests in the United States involving parents who have falsified their addresses to get their children into better schools.

International boundaries are common scenes of conflict and therefore often heavily fortified. In modern times, the border between France and Germany has probably been the most intensely fortified of all. World War I saw millions of soldiers killed on the battlefield in France. The Germans built an impressive defensive line called the Hindenburg Line, a portion of which was called the Siegfried Line. It must have worked, because the armistice was signed without Allied armies having taken any German territory whatsoever. World War I was a largely defensive war involving so-called trench warfare.

A common saying is that generals always prepare for the last war, and to some extent this was true of both France and Germany after World War I ended. The French built the famous and expensive Maginot Line between 1930 and 1939. This shouldn't be thought of as a line of forts or as a wall: the Maginot Line actually covered a zone between France and Germany that extended nearly twenty miles into French territory. Probably the French didn't consider the line impregnable, but they thought it would slow the Germans down sufficiently to allow mobilization of French forces. Of course, it didn't work at all. The Germans outflanked the wall and never confronted its elaborate fortifications.

Answer 30a: Later in the 1930s, after Germany had rearmed the Rhineland, completely contrary to the peace treaty it had signed, it built a defensive line of its own near the Rhine River. The Germans called it the Westwall but the Allies used the name "Siegfried Line" after the much shorter wall the Germans had built in World War I. It was intended to protect against an invasion from France. The Germans were never able to complete it to their satisfaction, but for years it didn't seem to matter. France had quickly succumbed to a

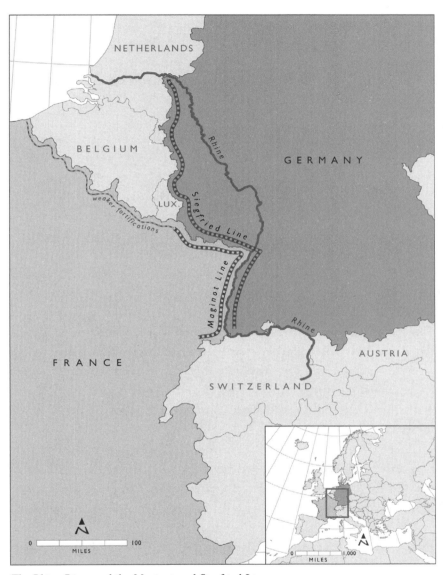

The Rhine River and the Maginot and Siegfried Lines

German invasion and offered no threat to the German border. After the D-day landings, when it might have mattered, the Germans undertook a major effort to complete and reinforce the Siegfried Line, but by that time it was largely obsolete. Air power and armor-piercing munitions made it all but useless.

In recent years, the Korean conflict, fought in the early 1950s, has been referred to as the "forgotten war." The *real* forgotten war in American history, however, was the War of 1812. Battle sites in New York State, such as Ogdensburg, Sackets Harbor, and Plattsburgh, don't come readily to the minds of even devoted history buffs, nor does the fact that parts of Maine were seized and held by British forces throughout the war. It seems that three combatants (the United States, Canada, and Great Britain) in the war all claimed victory . . . and all had good reason. The United States won some important naval battles, particularly on the Great Lakes and Lake Champlain, and they sacked and burned York, Ontario (modern-day Toronto). After the war was technically over, they won an important battle at New Orleans. The British were victorious elsewhere and burned Washington, DC. The Canadian militia, aided by Native American tribes and the British, repulsed three major invasion attempts by the Americans.

There were several reasons behind the US decision to declare war, and opinion in the country was far from unanimous. New England was opposed to the war and threatened to secede from the Union. Henry Clay of Kentucky was a leading proponent of war; he and others thought Canada would fall quickly into American hands. Their reasoning was that much of the population of upper Canada (near the Great Lakes and the upper St. Lawrence Valley) were formerly residents of the United States and would welcome repatriation (actually, this thinking was entirely wrong). The American frontier in 1812 was controlled by an alliance of the British and Native Americans, and the British maintained several forts on American soil. From Clay's perspective, this situation was intolerable.

The Treaty of Ghent, which ended the War of 1812, produced a curious reaction in both the United States and Canada: both countries were pleased with the result of the war and eager to enter into a peaceful relationship. **Answer 30b:** The result is that the US–Canadian border is the longest unfortified border in the world, possibly the longest such peaceful border in all of history.

Unfortified though it may be, the border isn't as easy to cross as it once was, and it's far more difficult to cross than it is to pass from one country of the European Union to another. When I was a mere lad, my parents and I crossed into Canada regularly for shopping or fishing and were waved through with little or no documentation needed. Now, passports or other entry documents are required. The last two times I crossed the border, there were major interruptions to my travel. My children were with me on the first crossing, and each one was asked where he or she was born. Each was born in a different state. This raised the proverbial red flag, since kidnapping children (particularly in custody disputes) and taking them across the border has become an important international issue. The next time I crossed the border from Canada to the United States, I had already passed through US immigration and customs at the Toronto airport, but when we landed in Chicago we were taken to a special facility and our baggage and persons were searched very thoroughly. We were not told why, but I doubt that it was an issue of smuggled maple syrup!

At the time the United States purchased Louisiana from the French, Spain owned all or parts of the current border states of California, Arizona, New Mexico, and Texas. In 1810, with **MEXICO'S INDEPENDENCE**, these areas became Mexican territory. Piece by piece, these areas became part of the United States, with the US–Mexican border finally established by the results of the Mexican-American War and the Gadsden Purchase. Issues associated with this border have occurred regularly over the last 150 years and currently are a major political issue in the United States.

Dr. G says: Mexico's Independence Day is September 16, and it's a big celebration. For reasons unknown, Americans seem to think it's on May 5 (Cinco de Mayo), an important day in Mexican history (when Emperor Maximilian was ousted) but not as big as September 16!

Although some current facts concerning the border are hard to ascertain, there are three clear points: (1) at least 10 million people now resident in the United States have entered the country illegally from Mexico, (2) while border patrol agents and other **SECU-RITY MEASURES** have been increased recently, it's the border region rather than the border itself that has been made more secure, and (3) Hispanic culture existed in this region for more than two hundred years before its residents became citizens of the United States by virtue of the territory ceasing to be part of Mexico and becoming part of the United States.

Dr. G says: An important part of the political rhetoric in the United States involves exactly this issue. It's nearly impossible (or would be downright lucky) to stop people as they cross the border and a lot easier to capture them later on. Americans who live very near the border aren't satisfied with that situation.

Political groups in both Mexico and the United States argue that the states of California, Texas, New Mexico, and Arizona should be returned to Mexico. While it's hard to imagine any circumstances in which this idea would be seriously considered by the United States, it is precisely this issue that was a force in bringing the United States into World War I.

Answer 30c: In 1917, the German foreign secretary, Arthur Zimmerman, sent a coded telegram to the German ambassador to the

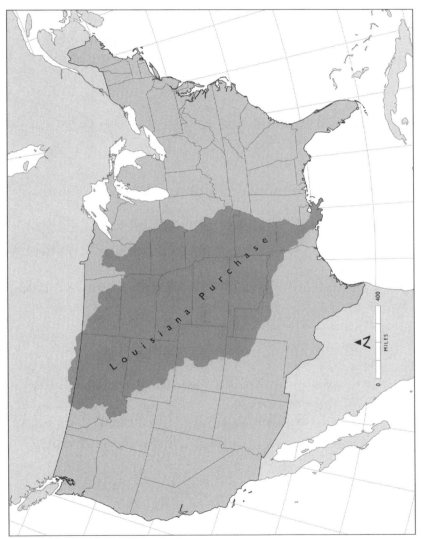

The Louisiana Purchase and Current US State Borders

United States instructing him to contact the Mexican government with a proposal. Germany proposed an alliance with Mexico that would feature a Mexican declaration of war against the United States in exchange for the return of **TEXAS**, **ARIZONA**, and **NEW MEXICO** to Mexican rule. The British intercepted and decoded the telegram and released it on March 1, 1917. An outraged United States declared war on Germany on April 6.

> Dr. G says: Why the Germans didn't offer California in this proposal is certainly puzzling to me.

The modern German state emerged first as the German Empire in 1871. At that time, Germany's borders encompassed more territory than they ever would again. Even so, territory that Germans considered their historic homeland was still outside the German boundary, which proved ultimately to be a cause of both world wars as the Germans were intent on expansion. The German Empire collapsed at the end of World War I, and the ensuing peace treaty caused major shrinkage in the German boundaries.

Even before Germany's defeat in the Second World War, the western Allies (at the Potsdam Conference) agreed that Germany would shrink even more. **Answer 30d**: The eastern boundary was established at the Oder River. The line was formally called the Oder-Neisse Line. The Neisse is a tributary of the Oder (and is called the Lusatian Neisse to differentiate it from other tributaries with the same name).

GEOGRAPHY
OF SPORTS

Question 31a: Of the forty-eight contiguous states, which is the largest in area that doesn't have Major League Baseball, NFL football, or NBA basketball teams?

Question 31b: What is the largest city in the United States that doesn't have an NFL franchise?

Question 31c: Where was basketball invented?

Question 31d: Who was the most prolific scorer in basketball history and in what state was the college for which he played?

Question 31e: Who has won the most NCAA basketball tournaments?

Question 31f: What is the westernmost member of the Ivy League?

New geographies unfold all the time, and it's unfortunate that geography departments in major universities are so slow to study them. The geography of retirement, for example, is at least as viable a field of geographic study in the United States and Canada as industrial/economic geography was a generation ago, but it's largely ignored. Similarly, athletics continues to grow in importance at many different levels—especially on college and university campuses—yet it's a rare geography department that includes anything associated with sports in its curriculum. Recent textbooks in geography devote considerable space to shopping centers but virtually nothing to the migration of the elderly and our preoccupation with sports.

It's really hard to say which is the bigger enterprise: professional or college sports. Professional sports franchises are worth billions of dollars, but the sports complexes on many campuses are breathtaking. College football bowl games have become so numerous that it's hard to keep track of them. And the line between professional and amateur sports is barely visible. Colleges devote considerable attention to keeping their athletes in school . . . and hoping they graduate.

Geographers developed a set of ideas that explained the size, locations, and functions of central places—usually urban centers. One idea in particular, "threshold and range," is useful in understanding why some places have big-time sports activities and others didn't. *Range* means how far people, on average, would travel (in this case) to a sporting event, while *threshold* poses the question of whether there are enough people living within the range to support the event. New York City, for example, has enough people living within a reasonable distance to support almost any athletic event. It's hardly surprising that New York has had two or three Major League Baseball teams for more than a century as well as two National Football League franchises. Elko, Nevada, on the other hand, is unlikely to get either Major League Baseball or an NFL franchise.

The location of major sports teams never has corresponded exactly to this geographic theory (called "central place theory"). The early National Basketball Association (NBA) had small-town teams that were initially quite successful. Syracuse and Rochester, New York, both had franchises, and the Syracuse Nationals won the NBA title in 1954, beating the Fort Wayne Pistons in the final series. The NFL's Green Bay Packers are an incredible anomaly, a small-town team that has lasted and lasted—in fact, it has achieved legendary success. Two other anomalies come to mind. Penn State University and Notre Dame have both achieved greatness in varying sports, most notably in football. Both schools' football stadiums are filled to overflowing on home game weekends in the fall, despite not being centrally located. Now, however, television has partially eliminated the effect of threshold and range. People can watch, and often pay for the privilege, thousands of miles from the actual game site.

Despite the influence of television, a threshold population is still needed to support the expense of operating a program and stadium. For this reason, there are no major sports franchises or even large college programs in lightly populated states. **Answer 31a:** The Dakotas and Wyoming, for example, lack such operations, as does Montana, the largest state in area among the forty-eight conterminous states to have no major league sports. **Answer 31b:** On the other end of the spectrum, it's surprising that Los Angeles, the second largest US city, lacks an NFL franchise.

Recruiting players for college sports is key to athletic success. From time to time, stories appear in the newspaper and sports magazines about supposed "geographic studies" of the prime recruiting areas for athletes in different sports. The trouble with such studies is that they don't take into account normal expectations. California and Texas, for example, by virtue of their population sizes, are likely to produce more athletes than Wyoming or Vermont. We also would have to take into account the propensity of players to stay reasonably close to home. It's not surprising that Michigan gets athletes from Ohio . . . it's right next door. It *would* be surprising if Boston College were getting significant numbers of recruits from Washington or Oregon. This kind of problem is meat and potatoes for the professional geographer! One common approach is the "gravity model," a simple formulation that takes into account population size and distance. Were we to apply this approach to different colleges and universities, we could identify recruiting "hot spots," and while they're at it, geographers could also plot recruiting routes that would minimize travel costs for athletic departments. It's certainly not the purpose of this book to report the results of this hypothetical research, but I can divulge an early finding: recruiting hockey players is easier in Canada than in Mississippi!

A complete geography of sports probably should take into account climate. Traditionally, you can play a lot more baseball in Southern California than you can in New England, and football offenses that require a lot of ball handling and exchanges can be run better in warm, dry climates than they can in Pennsylvania or Ohio. To a considerable extent, however, climate has been neutralized by covered stadiums or by fields with artificial playing surfaces. **Answer 31c:** In 1891, however, James Naismith was trying to cope with climate and find a way to keep his students physically fit during the long winters in Springfield, Massachusetts. He nailed a peach basket to the wall and thereby invented the sport. You probably got this answer correct, but what you probably didn't know was that Naismith was born in Canada (which is still south of Detroit).

Informal geography includes the biggest, longest, and most exceptional things. In American sports, despite thousands of competitors, it's still possible to find the single athlete in each sport who stood out above others: Babe Ruth in baseball, Jim Brown in football, **WAYNE GRETZKY** in hockey, Michael Jordan in basketball (who was a geography student at North Carolina). Despite these names, there's an almost forgotten basketball player who was the greatest scorer in the sport's history. He averaged more points per game than anyone in college basketball history and **TWICE SCORED OVER 100 POINTS** in a game. **Answer 31d:** Bevo Francis is the record holder, and he played for Rio Grande College. Give yourself a lot of extra credit if you knew that Rio Grande College is in Ohio!

Dr. G says: I have properly included Gretzky under the heading "American sports" because he finished his career with an American team.

In the late 1930s, legendary Kansas basketball coach "Phog" Allen started what's become known as "March Madness," the annual

Dr. G says: Francis scored more than 100 points on at least three occasions, including the all-time high of 116. Some of these scores weren't allowed in the record book, however, because Rio Grande played a schedule that included junior colleges.

National Collegiate Athletic Association (NCAA) basketball championship. The format of the tournament has changed several times over the years and now involves more than sixty teams playing at a number of venues around the country. A number of colleges are well known for their basketball programs and for their success in the tournament: Kansas, Kentucky, North Carolina, and Duke come readily to mind. **Answer 31e:** The team, however, that has won the most tournament titles by far (14) is UCLA.

When the Ivy League is mentioned, we think of academic excellence and prestige rather than sports. In fact, however, the colleges that now make up the Ivy League dominated collegiate football for many years. Yale in particular won more collegiate football games than any other school until its record was recently passed by the University of Michigan. The Ivy League insisted on maintaining admissions standards and applying scholarships on uniform criteria to all of its students.

Seven of the eight Ivy League schools were established in the colonial period of the United States: Harvard, Yale, Dartmouth, Brown, Columbia, Pennsylvania, and Princeton. The eighth is an outlier in three ways. **Answer 31f:** Cornell wasn't established until 1865, it's the westernmost (located in Ithaca, New York), and it was co-ed from its inception.

GEOGRAPHY AND
RELIGION REVISITED

Question 32a: Where is the Syro–Malabar religion found?

Question 32b: What country in the Caribbean did Emperor Haile Selassie visit in 1966?

Question 32c: What religious group was a significant force in colonial America but today is a relatively small denomination?

Question 32d: In what three countries are Shiite Muslims in the majority?

My grandmother believed that it wasn't polite to discuss religion or politics in her living room. Her idea seems to have taken hold among students in the college classroom as well. Most of my students weren't raised in a particular religion, and a surprising number know very little about religions in general. Electoral politics also seem to be out of style, with a surprisingly small percentage of students registering or voting in elections. Both religion and politics fall under the general headings of ideology: the core of both is belief, and those beliefs can't be tested scientifically. Beliefs are a type of idea that spread over time and space and therefore are very much within the domain of geography.

For those who've been raised with some religious background, it seems reasonable to include religion in a geography curriculum.

For others, however, it may seem to be offering an ideological perspective in a class that purports to be scientific. There's nothing ideological, however, about the influence of religion on European exploration, on numerous historical wars, on migration to the new world, and on contemporary political issues such as gay rights and abortion. Justice and legal systems (our reliance on guilt or innocence, for example) and matters concerning marriage and divorce are based ultimately on religious ideology. The economy of places, too, has long depended on the existence of holy places and pilgrimages. Places like Mecca, Lourdes, and Santiago de Compostela would be very different without the economic impact of religion.

Religion and politics have been unified throughout most of human history (and probably prehistory). Only in modern times have there been attempts to separate them. This creates a definitional problem: what exactly is a religion and how does it differ from a political belief? Most people when confronted with two belief systems (take Buddhism and communism, for instance) will identify one (Buddhism) as a religion and the other (communism) as a political belief. We generally think of a religion as having an element of worship, but the Four Noble Truths, which are at the core of Buddhism, don't mention the idea of worship or God. The Buddha himself was asked directly during his Deer Park sermons in Varanasi, India, "Whom do we worship?" He was silent. Meanwhile communism has featured scripture in the form of the *Quotations from Chairman Mao-Tse Tung* and the writings of Lenin, as well as figures who have at least approached divinity in the eyes of their followers: Mao himself and **KIM IL-SUNG** of North Korea are examples.

> Dr. G says: As testimony to the point made here, consider that Kim Il-Sung was named "Eternal President" after his death.

Religion over the past two millennia or so has been involved with one of the world's greatest processes of change. During the spread of universal religions they replaced preexisting religions and often adopted earlier religious practices as part of their own rituals or customs. A universal religion is one that actively seeks converts. The three major universal religions are Buddhism, Christianity, and Islam.

Ideas, such as religions, that spread over space and through time (and that's what the universal religions intend) inevitably confront a geographic reality: the character of places influences ideas that come to them. When Christianity began to spread through the Roman Empire, it confronted the reality that Roman customs were strongly embedded . . . and quite different from the customs of the Middle East, where Christianity originated. The Saturnalia was a major event in Rome and had existed for over two hundred years before Christianity. It was celebrated at the time of the winter solstice and lasted about a week. Christianity substituted Christmas for the Saturnalia. There's no scriptural basis in Christianity for celebrating Christmas on December 25—it's a present from the pagan Roman Empire.

The farther ideas spread, the more likely they are to undergo change. Christianity spread over a considerable area and, although it had both local and overall central administration, the sheer size of its domain implied that alterations in the religion were likely to be made at the edges of the pattern of spread. The first major split that has lasted to the present occurred in the form of the Great Schism of 1054. Christianity was split between a western branch that accepted the primacy of the Bishop of Rome (Roman Catholics) and those that denied that primacy (Orthodox Christians). An immediate cause of the split was a one-word change to the basic creed of Christianity, the Nicene Creed, allegedly made by Charlemagne, the Holy Roman Emperor. The Orthodox Church wouldn't accept a change to the basic creed made without its consent.

This eleventh-century split in Christianity was deep, enduring, and bitter. Attempts at reconciliation over the centuries have not been successful. There were, however, some eastern church groups that recognized the supremacy of Rome but otherwise retained the traditions and practices of the eastern church. One common name for these groups is *Eastern Rite Christians.* **Answer 32a:** One of these groups is based in Malabar, India. The Syro-Malabar Church is one of the more interesting branches of Christianity due to its age (one of the earliest churches) and because it's believed that its founders were Jews who believed in the divinity of Christ. Some of the practices of the church contain elements of both Judaism and Hinduism.

More than four centuries later, the Protestant Reformation split the Roman Catholic Church, this time along north-south lines. Martin Luther is usually given credit for initiating the Reformation in 1517. Although the initial rift was based on the corruption of the papacy, Luther's split differed from the Schism of 1054 in that it stressed basic changes in doctrine rather than only issues of papal authority. This fracture of Catholicism occurred largely as a geographer would have predicted: the Christian areas closest to Rome remained Roman Catholic, while those farthest away became Protestant. A notable exception is Ireland, an area about as distant from Rome as anyplace in Christendom at the time. Protestantism flourished when it unified with the national aspirations of the populace. In Ireland's case, Reformation was generally felt to be foreign and, especially, English!

The New World is sufficiently distant from the Old that we would expect religious beliefs to have undergone significant change as people and religious institutions migrated. In some cases, new religious groups developed: the Mormons are an example discussed in an earlier chapter. One of the more interesting groups to form is the Rastafarians, who began in Jamaica in the 1930s. The religion centers on Emperor Haile Selassie of Ethiopia, who's considered divine by many Rastafarians. **Answer 32b:** Selassie visited Jamaica on April 21, 1966, a day celebrated as Grounation Day by Rastafarians.

The United States and Canada became the refuges for many small Christian denominations, some of which disappeared entirely from their European homelands. Anabaptist groups such as the Hutterites (mostly found in the Canadian prairie provinces) and Amish (originally found in Pennsylvania but now distributed much more widely) are examples. **Answer 32c:** These groups are both pacifist, as are the Quakers, or Society of Friends, who were an important religious group in colonial America but have now become a tiny minority. William Penn, the founder of the colony of Pennsylvania, was a Quaker. Despite the decline in the influence of the Quakers, our only Quaker president served in the twentieth century: Richard Nixon.

Islam, the most recent of the universal religions, became dominant in areas that had previously been Christian and spread from the Atlantic to the Pacific as well. The principal division in Islam is between Sunni and Shia Muslims, a difference that is now well known to both my students and the American public. The split between the two groups concerned the successor to the Prophet Mohammed. Shiites believe **MOHAMMED** clearly named who was to follow as leader, but Sunnis thought that an election of the new leader should be held. The minority Shiites look for a leader who will restore Islam to its former glory. **Answer 32d:** Shiites are in the majority in Iraq and Bahrain and are an overwhelming majority in Iran.

Dr. G says: There has not been a universally recognized leader of Islam for more than a thousand years, so in a sense the original reason for the split is a moot point.

THE NORTH

Question 33a: Where did the word berserk *originate?*

Question 33b: What is the most recently settled country in the world?

Question 33c: What is the only NATO (North Atlantic Treaty Organization) member without a standing army?

Question 33d: The capital city of which country is farthest north: Denmark, Norway, Sweden, or Finland?

Question 33e: What two countries fought the Winter War in 1939–40?

Question 33f: Who was the first European to discover Alaska, and what was his nationality?

For about three hundred years, from 800 to 1100 AD, a relatively small group of people in the north of Europe transformed their society from dependence on farming and trading to reliance on raiding. They terrorized and pillaged areas in Europe, the Mediterranean, and parts of Asia and explored lands previously unknown to Europeans. These people, whom we lump together as Vikings, accomplished things that, in retrospect, seem almost impossible. The fiction written about these people is probably more believable than the truth, but we need to start our discussion of them by dismissing two widely held beliefs. First, the Vikings never wore helmets with horns on them in battle! This information may not seem particularly valuable, but it's a way of testing source material about the Vikings: in my experience, sources that are illustrated with horned helmets are not reliable. Second, there's no evidence that Vikings disposed of their dead by placing them in boats and setting the boats on fire. They did, however, bury their boats . . . and the dead with them!

Whenever a group of people achieve great military victories over a long period of time, it's reasonable to assume that they have weapons that give them superiority. For the Vikings, their boats were such a weapon. The Viking longships were very strong craft, with true keels, and were made of split oak planks. They were light enough so they could be carried by their crew, and they were perfect for raiding. A landing could be made by beaching the boats, then the raiding party could be evacuated without need to turn the boats since either end could serve as the bow. Their speed in the open water allowed them to escape easily from other ships. The boats, however, weren't intended as naval craft but as cargo ships and troop carriers.

The Vikings were feared wherever they roamed. They apparently were somewhat larger in stature than other Europeans at the time, and their ability to bring forces rapidly ashore and move inland from a landing site meant that they often had surprise on their side. **Answer 33a:** Some Viking warriors would fight in a trancelike state and uncontrollable frenzy. They were called "berserkers," from which we get the English word *berserk*.

Vikings, especially those from western Norway, explored and settled the Faroe Islands, which are north of Scotland, as well as Iceland, Greenland, and Newfoundland. (The Newfoundland settlement wasn't permanent, as it was too far from the Viking homeland.) **Answer 33b:** Iceland was probably the last country in the world to be settled. Permanent settlement dates to the ninth century; only **NEW ZEALAND** is a rival for the title of "most recently settled country." Iceland, however, with a population of about 320,000, makes New Zealand's population of 4.4 million seem huge in comparison. Iceland has been prominent in the news during the last few years because of its volcanic eruptions. The country, noted for its geothermal energy, is in a highly active volcanic and tectonic region. The volcanoes have carried ash to high levels of the atmosphere, resulting in the cancellation of some transatlantic and European flights. **Answer 33c:** Iceland is a member of NATO (North Atlantic Treaty Organization) and hence a military ally of the United States and Canada, but it's the only NATO member without a standing army.

> Dr. G says: Evidence that New Zealand was settled in the second century AD has been called into question, and the first Maori may have arrived as recently as the thirteenth century.

The Vikings travelled to the Mediterranean and Middle East as well as to North America. They established a kingdom on the island of Sicily and were the immediate forebears of the people we know as Normans, who successfully invaded Great Britain in 1066. The Vikings also used their boats in the rivers of Russia, especially the Volga, and established a settlement at Kiev, now the capital of Ukraine. In fact, *Rus* was the name Vikings gave to these eastern European lands. Archeological evidence shows that the Vikings even reached Baghdad, which at the time was the center of the Arab-Islamic world. For reasons we can only speculate about, the Vikings highly prized silver. The Viking siege of Paris was lifted by payment of silver, and it's quite possible that their voyages destined for the Middle East were motivated by pursuit of this precious metal.

The Northern European countries (except Russia) have long been oriented toward the sea. In the past, they depended on trade and fishing as mainstays of their economies. Their capitals are all seaports and are the subjects of numerous geographic trivia questions. **Answer 33d**: Reykjavik, Iceland, is the most northerly of all the capitals, but of those on the European mainland, Helsinki, Finland, is the most northerly. There's a strange twist to this question, which occasionally appears in trivia games: What's the capital of the most northerly land in the world? The answer turns out to be the most southern of the northern capitals, Copenhagen, Denmark. The **MOST NORTHERLY LAND** in the world is a small island off the coast of Greenland: Danish territory!

> Dr. G says: Until recently Canada claimed to be the northernmost country, but more reliable surveying now shows that Greenland is farther north by only a few hundred feet!

Throughout history, wars were usually fought in the summer when movement across open country was easier. George Washington, for example, had his troops in a winter encampment at Valley Forge, Pennsylvania. This doesn't mean that fighting ceased entirely in the winter: Washington's crossing of the Delaware to attack the British near Trenton on Christmas Eve 1776 is an example. **Answer 33e**: In the Winter War between Finland and the USSR, fought in the winter of 1939–40, the geography of Finland dictated the timing of the war. The world had watched as Hitler had unleashed his blitzkrieg against Poland in September 1939. The Germans' rapid deployment of tanks changed battlefield tactics. The Soviet Union, fully expecting an attack from Germany, moved to secure defensive positions around Leningrad (earlier, and yet again, St. Petersburg) and in the Baltic. They demanded and received the capitulation of the Baltic countries (Estonia, Latvia, and Lithuania) and sought also to occupy Finland. Finland had been a vassal state of the Russian Empire and had obtained its independence only a few years earlier. Finland refused to be occupied. The USSR invaded in the winter (in part) because its tanks

Norway, Sweden, Denmark, and Finland

would be effective only when the many lakes and swamps of Finland froze over. The Finns resisted a full-scale invasion and held out until spring. The ensuing peace treaty favored the USSR, but the Finns maintained their status as an independent country.

The Nordic countries have achieved, by somewhat different routes, high standards of living for their populations. Denmark has been characterized as the "world's happiest country," and recent reports show that Norway's per capita income is the highest in the world. One thing they're not happy about, despite economic prosperity, however, is the lack of sunlight during their winters. Those who can afford it take winter vacations in southern Europe or other locations where there are longer days! In compensation, the long summer days are spent by many in the rustic cabins scattered through the forests and lakes of Norway, Sweden, and Finland.

Most of us know that Alaska was once owned by Russia and was purchased by the United States in 1867 through negotiations conducted by US Secretary of State William Seward. Although our history books remind us that Alaska was called "Seward's Folly," public opinion and the US Senate were very much in favor of the purchase. The US House of Representatives, however, balked at paying the outrageous purchase price of 2 cents per acre! Eventually, the House did approve the purchase by a substantial margin. But how did the Russians get Alaska in the first place?

The first European to visit and claim Alaska was captain of a Russian ship and known to his crew as "Ivan Ivanovich." **Answer 33f:** His real name, however, was Vitus Bering, of Danish nationality, and many things Arctic bear his name, including the Bering Sea. His expedition discovered Alaska in 1741.

CHAPTER 34

PRISON ISLANDS

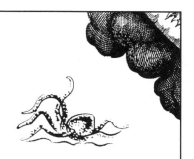

Question 34a: Where were convicts, originally banished to Australia, sent for committing a second serious crime?

Question 34b: Where was Al Capone imprisoned?

Question 34c: Where was Alfred Dreyfus imprisoned?

Question 34d: Where was Dr. Samuel Mudd, who was convicted of aiding and conspiring with John Wilkes Booth, imprisoned?

Question 34e: With what three islands was Napoleon Bonaparte associated?

Increasing connectivity between places occurs for various beneficial reasons. In the industrial age, good connectivity meant the ability to assemble raw materials at minimum cost and then have easy access to market for finished manufactured goods. In the contemporary world, good connectivity also means we can more easily visit places for recreation and for visiting friends or family.

Prisons, however, are usually located to minimize connectivity. Placing them in remote locations seems like a good idea—the more remote, the better. Although some famous prisons, like the Tower of London and the Bastille, are **CENTRALLY LOCATED**, the logic behind their existence is a bit puzzling. Why pay to house and guard someone who you could execute or remove from your locale by banishment or by what the British called "transportation"? Sending prisoners to distant locations not only saved on the cost of prisons, but the prisoners could be put to work on some worthwhile project with minimal labor costs. There was also a moral benefit to this practice: when prisoners were moved to a distant location, they were "removed from temptation," according to those who advocated banishment.

> Dr. G says: Although most modern prisons are located in rural, out-of-the-way places, some are found in central cities. One of the most famous is San Diego's high-rise prison, which is surrounded by expensive condominiums and luxury hotels.

Georgia, the last of the thirteen original American colonies to be established, is a good example of the use of banishment. James Oglethorpe, supported by Parliament, established the colony of Georgia as a place for the "worthy poor" of Britain. By the late 1730s, prisoners were being transported to Georgia, many for having unpaid debts (this was clearly before the invention of the credit card!). The North American colonies ceased to be a destination for British pris-oners after the American Revolution, but Australia took up the slack.

Possibly the most famous prison colony in the world was the one that wasn't at Botany Bay, Australia. However, that's where Captain Cook first went ashore in Australia. "Botany Bay" wasn't his first choice as a name, but the flora that surrounded the bay was so unique and interesting to botanists that he changed his mind. The British wanted a penal colony there, but the immediate area around Botany Bay wasn't suitable, and the colony was moved to another bay to the north. The British seem to have been stubborn about this matter, however, and the colony was commonly referred to as Botany Bay even though that's not where it was!

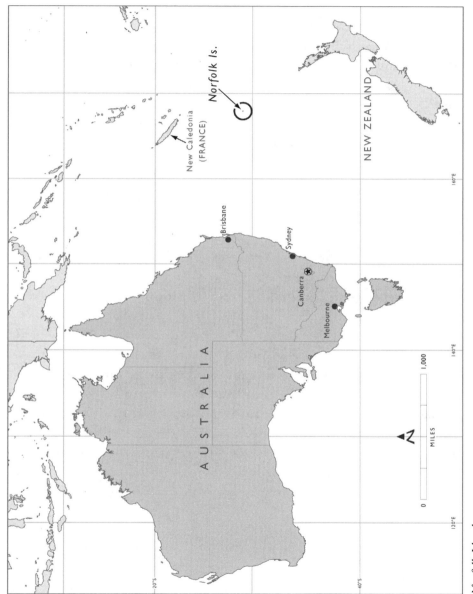

Norfolk Island

Australia is about as far from Great Britain as possible and for that reason alone seemed like a great place to have a prison colony. A problem arose, however. **Answer 34a:** What do you do with prisoners who've been transported to Australia because of their criminal behavior and then reoffend? The answer, it turned out, was to send them to Norfolk Island, a small uninhabited land about equidistant from Australia, New Zealand, and New Caledonia. Although the island had once been settled by Polynesians, they had abandoned their settlement by the time the British (again, in the person of Captain Cook) arrived. Twice a penal colony was established there for Australian prisoners, and it was twice abandoned. Those who stayed on Norfolk Island, however, were descendants of some of the crew of the *Bounty* who had been previously living on Pitcairn Island!

Geographers interested in place names know that a name beginning with *Al* is often originally an Arabic word, adopted by the Spanish during the centuries the Moors occupied Spain. This is the case with the most famous island prison in the United States: **ALCATRAZ**. Although Alcatraz was only in use as a federal prison for civilians for about thirty years (it had earlier been a military prison) it housed some of the most infamous and dangerous American criminals. It was a place for the "baddest" of the bad. Most of the inmates had been in other prisons previously and had committed crimes while imprisoned. Probably its most famous prisoner, however, didn't fit this profile. **Answer 34b:** Al Capone was convicted only of tax evasion and originally was imprisoned in Atlanta. He was sent to the isolation of Alcatraz to keep him from running his mob organization while in prison.

Dr. G says: The word *Alcatraz* refers to sea birds. Exactly which bird, however, is debatable, as the word is now considered archaic Spanish.

Answer 34c: Being sent to prison when you're unquestionably inno-
cent is a terrible fate, even more so when the prison is Devil's Island.
That was the fate of Alfred Dreyfus, a French Army captain falsely
convicted of selling military secrets to the German embassy in Paris.
Devil's Island is the northernmost in a group of three small islands
off the coast of French Guiana, on the Caribbean coast of South
America. Collectively, the islands are called the **"HEALTHY
ISLANDS"** since it was felt they were a lot healthier than French
Guiana itself. Dreyfus was kept in solitary confinement there for two
years. Eventually he was exonerated and returned to the army. He
served as an artillery officer in World War I, ironically using the very
weapons whose secrets he had been accused of selling to the Germans.

> Dr. G says: Devil's Island may be healthier than the
> mainland of French Guiana, but after having spent sev-
> eral hours on the island, the difference was not markedly
> apparent!

The Dreyfus affair, made famous by an essay written by Émile
Zola, was more than a routine injustice. The case against Dreyfus
seems to have been based ultimately on the fact that he was Jewish.
Jewish leaders in Europe questioned how a country as advanced
as France could have such embedded anti-Semitism. The Zionist
movement—an attempt to restore the homeland of the Jews—was
given a significant boost by the Dreyfus affair. Jews, the Zionists
argued, could be safe only in their own country. A terrible twist to
this story is that the rulers of **VICHY FRANCE** during World

> Dr. G says: Vichy France was essentially a German pup-
> pet state established after Germany conquered France in
> 1940. The leaders were French, but many of them had
> been anti-Dreyfus and anti-Semitic.

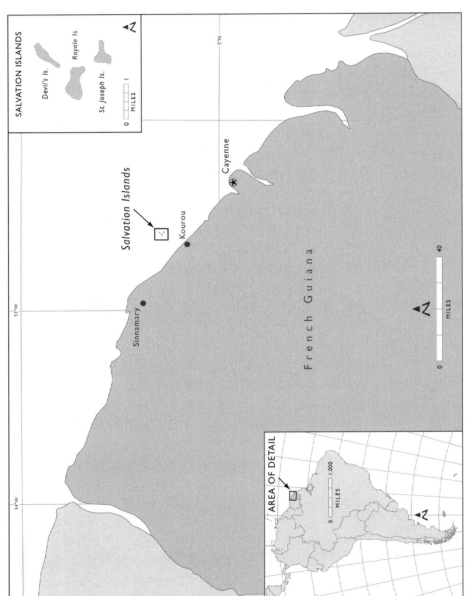

Devil's Island

War II turned a granddaughter of Dreyfus over to the Germans. She was sent to Auschwitz, where she died.

Following the assassination of Abraham Lincoln by John Wilkes Booth in 1865, a southern Maryland physician and slave owner, Samuel Mudd, was convicted of aiding Booth, who had broken his leg while jumping to the stage at Ford's Theatre. Booth escaped to Mudd's house, where the physician set and splinted his leg. Mudd claimed that he hadn't known about the assassination and had treated Booth as he would any patient. There was evidence, however, that Booth and Mudd had met earlier, and Mudd inexplicably delayed reporting that he had treated and sheltered Booth even after word of Lincoln's death had become widespread.

Mudd, along with other alleged conspirators, was tried by a military tribunal. Of those found guilty, all were hanged except for Mudd, who was sentenced to life in prison. **Answer 34d:** Dr. Mudd was imprisoned at Fort Jefferson, in the Dry Tortugas island group off the coast of Florida. The specific island he was on is called Garden Key. In 1869, President Andrew Johnson pardoned Mudd. In part, the pardon was given in response to Mudd's work as a physician at the prison during a yellow fever epidemic.

Napoleon Bonaparte changed the political geography of Europe. His victories threatened to bring most of Europe under French rule, while his defeats, especially at Waterloo and in Russia, have become legendary. The Napoleonic Wars, fought between 1800 and 1815, should properly be called the First World War since they involved battles from Belgium to Moscow and from Egypt to the Caribbean. One of his worst defeats, the virtual eradication of his forces in Haiti, was mentioned earlier. The defeat in Haiti was especially demoralizing since it was not at the hands of highly trained Prussians but by former slaves.

These wars saw the introduction of new ideas that changed the way wars were fought. France's mobile kitchens allowed a better means of

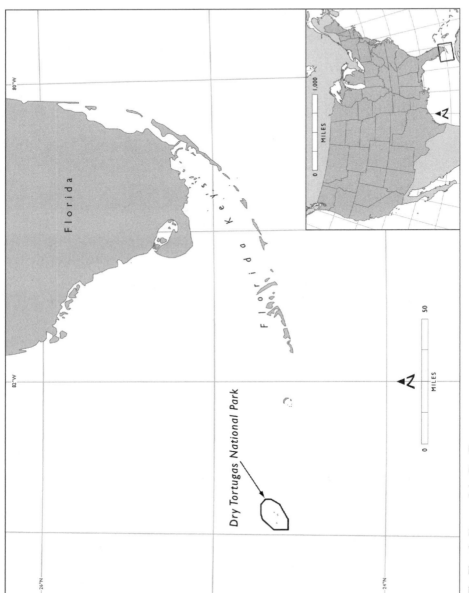

The Florida Keys and the Dry Tortugas

provisioning troops, and the French army used the 1810 invention of Nicolas Appert: canning. Appert, incidentally, advocated canning most anything that could be eaten. He even canned whole animals! France, under Napoleon, also used large-scale conscription to build his army. Prior to that time, wars had been fought principally by professional soldiers. The Napoleonic Wars thus had an effect that particularly interests population geographers. Following these wars, France experienced the first postwar baby boom as the drafted soldiers returned to their families.

Napoleon was born on the island of Corsica, which had fallen under French control only about a year before he was born. His parents were Italian nobility. As a young boy he went to France and was mocked at school because of his heavy accent. He aspired to enter the British navy but was admitted to the French military academy and became an artillery officer. His career was punctuated by both military and political achievements, and eventually he became crowned emperor of France.

The huge army he led to Moscow was virtually destroyed by the Russian winter and the "scorched earth" practice of the Russians. The lack of food for the French horses was key to the Russians' success. Shortly after, a coalition of Napoleon's enemies invaded France. Under the peace treaty, Napoleon abdicated and was exiled to the island of Elba.

Elba, however, didn't prove to be an island prison for him. In fact, Napoleon became emperor of Elba and introduced changes in the infrastructure of the small island. Apparently acting on rumors that he was to be removed from Elba, Napoleon returned to France. Within the famous "hundred days" of his escape, Napoleon managed to regain control of the French army, then lose the Battle of Waterloo to the British and Prussians. Following this defeat, Napoleon was sent to yet another island prison, St. Helena in the South Atlantic Ocean, one of the most isolated spots on earth. The British always feared that sympathizers would try to rescue Napoleon and return him to France. St. Helena was therefore heavily guarded

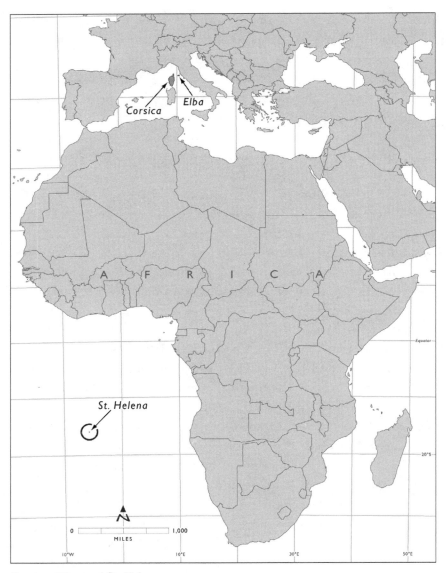

Corsica, Elba, and St. Helena

both on land and by the British fleet. **NAPOLEON DIED** on St. Helena in 1821.

> Dr. G says: Napoleon is buried in Paris, at Les Invalides, a hospital he established for war veterans. It's better known today, however, as the French War Museum.

Answer 34e: The three islands that were important in Napoleon's life, therefore, were Corsica, Elba, and St. Helena.

CANALS

Question 35a: What route was called "the Highway to India"?

Question 35b: Where is Gatun Lake?

Question 35c: When the Göta Canal was built in Sweden and the Erie Canal in New York, both in the first half of the nineteenth century, what bodies of water did each canal connect?

Question 35d: Where is the longest canal?

Question 35e: The opening of what canal allowed shipping to bypass Niagara Falls?

When we read of the Dark Ages in Europe, the period following the collapse of the Roman Empire, we're dealing principally with a sharp decrease in connectivity among places. Roads, some of them built by the Romans themselves, became dangerous to travel without armed escorts. Sea lanes, once kept open by the Romans, were subject to piracy or, in the north, to the Vikings. Individuals lived in relative isolation. Those with rare talents or interests were often not aware where others with similar interests were located. How does one study mathematics if one doesn't know where to find mathematicians? In other words, this was a period when one's personal geography was limited by lack of connectivity. Some short canals were built early in European history, but major canals didn't come on the scene in Europe until after the Middle Ages. When they did appear, they were an important contribution to improved connectivity. This contribution is often overlooked in today's world, replete with highways, trains, and airlines.

Canals that carry traffic and improve connectivity are the most interesting kind (a canal is technically any man-made excavation that carries water—even a common ditch can be considered a canal), but they were the downfall of my college students. Every semester the final exam would contain one or two questions about canals, and the students would miss them more often than any other questions. I was tempted to try the "Campus Center Cafeteria solution." Students who ate in the cafeteria loudly complained that the sandwiches they bought had wilted lettuce, so the cafeteria eliminated lettuce from the sandwiches (without, of course, lowering the price). I, however, left the canal questions in the exam and let the unprepared students wilt!

The Suez Canal was an enormous contribution to improved global connectivity. It dramatically shrunk the world at a time when European powers like Britain, France, and the Netherlands held important colonies in Asia. With the opening of the canal, connectivity between colonies and mother countries improved. A trip around the continent of Africa was avoided. Although there was previous shipping through the Isthmus of Suez, it required freight to be unloaded on the Mediterranean side, transported by land, then reloaded on the Red Sea side. That anything taking this much time and trouble was actually done is convincing proof of how valuable bypassing the African route was. The completion of the American transcontinental railroad in the same year the Suez Canal was opened substantially reduced the time it took to travel not only across America and from Europe to India but **AROUND THE WORLD**.

> Dr. G says: The public was caught up with this by the actions of newspaper reporter Nellie Bly and the fictional character she emulated, Jules Verne's Phileas Fogg.

Answer 35a: The Suez Canal was nicknamed "the Highway to India." Since India was the jewel in the crown of the British Empire, you'd certainly think that the British would be enthusiastic supporters of a canal at Suez, since it would reduce the distance to India by thousands of miles. In fact, the British opposed it and actually tried to interfere with the construction of it once it began. While the British were never great admirers of the French, their opposition probably was based largely on the fact that British shipping to the Far East and Australia utilized sailing ships, which couldn't be used in the Suez Canal. The British had invested heavily in clipper ships—very fast, square-rigged vessels. Their highly lucrative tea (and opium) trade with China depended on these ships. Within a decade after the opening of the Suez Canal, the clipper ship trade with China was entirely ended (although clipper ships continued to carry cargo between Great Britain and Australia and New Zealand for a few more years).

Ferdinand de Lesseps was the French developer of the Suez Canal. It was built at sea level, with no intervening locks. De Lesseps was forced continually to defend his project against severe criticism from all quarters. He was one of history's most determined men and upon completion of the canal became a great hero in France. His determination, however, was to prove his downfall when he took on his next project.

The Panama Canal may have been the greatest construction project ever completed. The French, again led by de Lesseps, attempted the job as if they were redigging the Suez Canal. Topographically and geologically, Panama is very different from Suez, and the contemplated canal was to be at sea level, something that at the time was

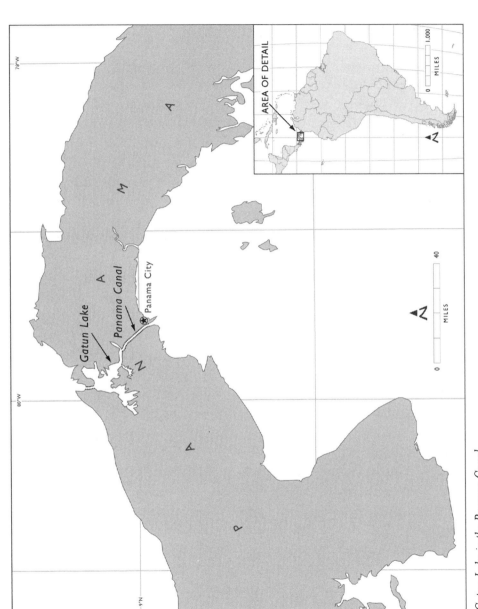

Gatun Lake in the Panama Canal

impossible for anyone to build in Panama. The French design, and their own construction plans, seemed to ignore the presence of the Chagres River, which ran along the route. The volume of water in the river was great, particularly during the rainy season. While de Lesseps could be characterized as "determined" in Suez, in Panama he was "stubborn." The French were defeated in their efforts, and de Lesseps was transformed from national hero to a disgraced man.

The ideal place to build a canal to connect the Atlantic and the Pacific was actually in Nicaragua, not Panama. When the Americans took over the job of building a canal after the French left Panama, Nicaragua was seriously considered as an alternate site. The presence of active volcanoes in Nicaragua (and their absence in Panama) may have been a decisive factor in the decision favoring Panama. Many American leaders feared that volcanic eruptions could destroy a Nicaraguan canal. Ironically, the Americans reconstructed the topography of Nicaragua in Panama. **Answer 35b:** Lake Gatun, in the middle of the Panama Canal, is the geographic analogue to Lake Nicaragua. The volume of water in Lake Gatun, which covers thirty-three miles of the canal route, is equal to the discharge volume of the Chagres River, which was dammed to form the lake. The Panama Canal is thus a hydrological balancing act with nature continually replacing the water lost in the canal operation.

Both the Suez and the Panama canals were late additions to the world. The canal-building era had begun earlier, in the eighteenth and early nineteenth centuries. By the time the canals across Suez and Panama were completed, famous earlier canals had already been made obsolete. An American example was the Erie Canal, which was completed in 1825. **Answer 35c:** The Erie Canal connected Lake Erie with the Hudson River and thus opened up the Great Lakes to a harbor on the Atlantic. The canal assured New York City primacy among American cities. By the time the Suez and Panama canals were built, most traffic on the Erie Canal had been replaced by railroads.

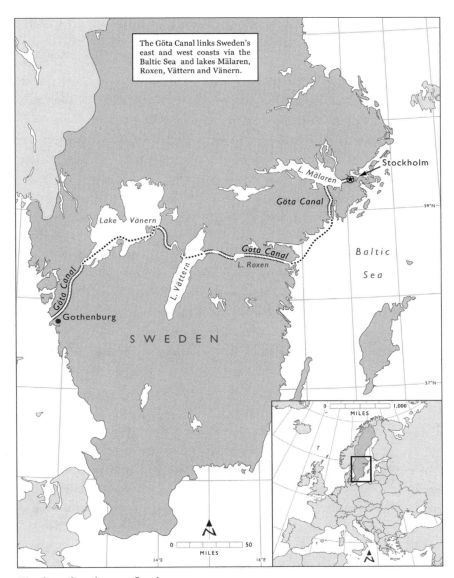

The Göta Canal links Sweden's east and west coasts via the Baltic Sea and lakes Mälaren, Roxen, Vättern and Vänern.

The Göta Canal across Sweden

A similar situation happened in Sweden with the opening of the Göta Canal. **Answer 35c:** The Göta Canal connected Gothenburg, on the North Sea, to the Baltic Sea on the east coast of Sweden. It's approximately the same length, and took about the same time to build, as the Erie Canal. Because it was built even later in the nineteenth century it was rendered obsolete by railroads sooner after its opening. European canals like the Göta are often used for recreation today and have become the focus of a relatively new business: recreational canal cruises.

The Chinese began building canals more than a thousand years before the Europeans. **Answer 35d:** China's Grand Canal is the longest canal in the world and, in simple fairness, deserves more fame and accolades than the Great Wall. Bits and pieces of it have been added and improved over the years, but the basic canal was completed around 600 AD. European and North American canals were built mostly to facilitate international trade and ocean shipping, whereas the Grand Canal was built primarily so that grain could be transported more easily from the richer farmlands of the south to Beijing, the capital.

Biogeography is the study of, among other things, invasive species. Often when a plant or animal is moved from its native habitat it can enter into an area where it has no natural enemies, and thus it threatens native species in its new habitat. This is becoming a larger problem because connectivity (that word again!) is increasing among most places in the world. Since connectivity is enhanced by canals, it follows that canals promote the transmission of invasive species.

One of nature's great natural barriers that impeded the movement of people and other species is Niagara Falls. The Niagara River is the outflow from Lake Erie to Lake Ontario, and the falls is the crowning feature of the river. A canal around the falls is an obvious way to improve transportation, but it's also a means by which species previously stopped by the falls can enter Lake Erie and the other

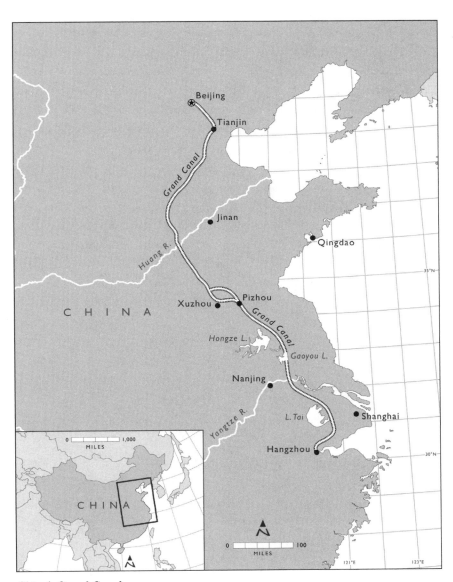

China's Grand Canal

Great Lakes to the west. Although several species have invaded the Great Lakes, the most famous is the lamprey. The lamprey looks like an eel, but it's not a true eel. It attaches itself to fish and feeds on them. **Answer 35e:** The canal often blamed for this invasive species is the Welland Canal. Although it bypasses Niagara Falls it's not the only means by which plants and animals can enter the Great Lakes; the Erie Canal itself provides a similar connection.

Federal and state agencies have attempted various ways to eradicate, or at least control, the lamprey population in the Great Lakes. The obvious solution is ruled out by the cultural difference between Europeans and North Americans. Although the lamprey has long been part of European cuisine, it's simply not eaten by people in North America. It's alleged to have been the favorite dish of King Henry I of England. His death has been attributed to overindulgence in a meal of stewed lamprey eels!

LAND AND
SEA BATTLES

Question 36a: In what river was the German ship Graf Spee scuttled?

Question 36b: What city was built on islands in a lagoon in the Adriatic Sea in the fifth century?

Question 36c: What famous battle started with the sinking of a miniature submarine?

Question 36d: In what state was the Civil War battle of Antietam fought?

Military geography was once an important subfield of the science of geography, both as a stand-alone curriculum and as part of political geography. Moreover, several of my professors, and later some of my colleagues, were military intelligence officers, a field for which the study of geography well prepared them. A few geography departments currently offer intelligence analysis as a career track. In the overall scope of military geography and intelligence analysis, the importance of individual battles seems small in the panorama of major conflicts, and some battles that seemed crucial at the time they were fought fade in significance as the passage of time gives us new perspectives. Some battles, however, are of special interest to geographers.

THE BATTLE OF THE ATLANTIC

Although there were many famous European battles in World War II, the Battle of the Atlantic served as a backdrop to all of them. The battle was notable not only for its area but for its duration as well: it lasted from 1939 through 1945. The objective in the Battle of the Atlantic was clear to both sides. If the United Kingdom were to hold on in its resistance to Nazi Germany, it needed to receive about one million tons of shipping per month at a minimum. The United Kingdom didn't produce enough food for its citizens and had long depended on its colonies and other countries. Germany believed it could deny that shipping and thus force the United Kingdom out of the war.

Upon the entrance of the United Kingdom into World War II in September 1939, the Germans sent three large surface ships into the Atlantic with orders to sink merchant ships. The German warships were built according to limitations set by the peace treaty ending the First World War and therefore were smaller than regular battleships (they were sometimes called "pocket battleships" and sometimes "heavy cruisers"). These, however, were among the first ships built with welded rather than riveted hulls and also used innovative diesel propulsion systems. As a result, they were lighter and could carry more armament. At this point, both the Germans and the Allies assumed that the Battle of the Atlantic would feature surface ships. The most famous of the three, the *Admiral Graf Spee*, was damaged in a battle in the South Atlantic in December 1939 and took refuge in the neutral port of Montevideo, Uruguay. **Answer 36a:** By international law, the *Graf Spee* was forced to leave Montevideo within seventy-two hours. Rather than allow his capture or loss of life among his crew, her captain scuttled **H I M** in the harbor just outside Montevideo, in the Río de la Plata.

As the Battle of the Atlantic continued, it became increasingly obvious that it would be mainly a battle between German submarines and Allied antisubmarine warfare. What emerged was a classic geo-

The Estuary of the Río de la Plata

D. G says: Strangely, some of the ships built by Germany during the period between the two world wars, including the *Graf Spee*, were referred to by masculine pronouns rather than the feminine pronouns conventionally used for vessels. If this attempted to make a point, it certainly didn't work!

graphic problem of controlling space. Submarines don't have much superstructure and therefore can't place lookouts in a position to see much of the ocean. Finding ships to sink, therefore, was their main problem. They successfully used land-based aircraft to spot merchant shipping with the location of the targets, then transmitted to the submarines via coded radio messages. To counter this strategy, the Allies attempted to destroy spotter planes and twice were able to capture German submarines with their analogue code machines intact.

The geographic problem for Allied shipping was how to arrange convoys of ships to minimize the chances that they would be discovered and attacked by submarines. The counterintuitive strategy that seemed to work best was to send ships in relatively small, lightly guarded groups. Although ships arrayed in these formations proved harder to find, the carnage created by the Germans remained high through most of the war. The Allies were, of course, able to keep the United Kingdom supplied and, by the end of the war, it was dangerous for a German submarine to surface almost anywhere in the Atlantic. In terms of ships sunk and lives lost, however, the balance sheet favored the Germans by a significant margin.

THE BATTLE OF LEPANTO

Euro-American civilization has a tendency to view the Islamic world as monolithic even though it's made up of many different peoples who speak many different languages. The Arab and Berber conquerors of Iberia, who threatened Christian Europe from the southwest,

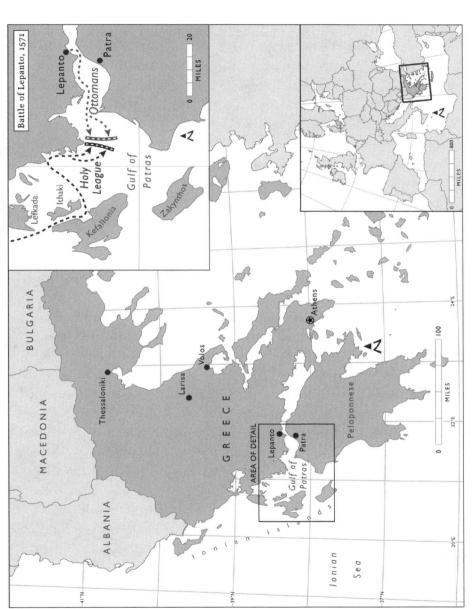

The Battle of Lepanto

had been expelled from Spain in 1492, but that didn't remove the possibility that an Islamic threat might come from elsewhere. By the sixteenth century, the Ottoman Turks who had captured Constantinople and destroyed the Byzantine Empire were threatening Europe from the east. In 1571, the Pope created a Catholic alliance that aimed at stopping the Turkish threat. A huge fleet was assembled to move against the Turks in the Eastern Mediterranean.

Strangely enough, the fleet was led by John of Austria, an area not known for its navy! John, however, was the illegitimate son of Charles V, arguably the most powerful of the Holy Roman Emperors. The largest contributor to the fleet was the powerful city-state of Venice, whose trade had been severely reduced by the Turks. **Answer 36b:** Venice, built in the fifth century on islands in a lagoon in the Adriatic Sea, had become a major naval power and had recently built craft that carried heavy cannons, something the Turks didn't have. Genoa also sent naval craft. Spain, unwilling to commit all its ships, sent over ten thousand soldiers to accompany the fleet. The papal fleet left from Sicily, and the Turks left from Lepanto. They met just south of Greece.

The Battle of Lepanto wasn't known for its tactics, strategy, or duration (it lasted a single day, October 7, 1571). It became famous as the last great battle between oar-driven ships, but some authorities argue it was the most important naval battle ever fought. The decisive victory of the papal fleet ensured that Europe wouldn't be overrun by the Turks. The Turks rebuilt their fleet and seized Cyprus, again attaining hegemony in the eastern Mediterranean. Islam, however, is a religion built on the fundamental principle of submission to the will of God. Major defeats are often interpreted as the will of God and therefore accepted. By 1580, the Turkish naval fleet was virtually nonexistent.

THE BATTLE OF PEARL HARBOR

This is the smallest of the battles considered in this chapter but nevertheless it has several aspects that make it of special interest to geographers. Perhaps first and foremost is that Pearl Harbor dem-onstrated the disappearance of the protective blanket of distance that had long sheltered the United States from attacks by an enemy. Even in the American Revolution the British were forced to fight a war far from their home bases, but later, when the United States became a substantial naval power, the protection offered by an ocean and na-val fleets on each side made the country all but impregnable. True, Hawai'i was a naval base 2,500 miles from the mainland, but the size of the fleet at Pearl Harbor and the air cover provided by the air bases on Oahu made it virtually invulnerable, particularly since the only likely enemy, Japan, was more than 4,000 miles from Pearl Harbor.

George Washington had warned the country to avoid becoming enmeshed in the wars in Europe, and many Americans continued to believe that should be the main principle of US foreign policy. In 1941, the United States remained an isolationist power. American commercial interests were clearly being affected by World War II, which had been under way for more than two years, and Congress had approved programs to aid the United Kingdom, but American public opinion favored staying out of the war. Underlining this iso-lationist sentiment was the fact that President Woodrow Wilson had been the prime force behind the creation of the League of Nations after World War I, but the US Senate had refused to ratify the treaty that would have made the United States a member of the League.

Answer 36c: The Pearl Harbor attack actually began with Japan's launching of miniature submarines. In the early morning hours of December 7, 1941, the minesweeper *Condor* sighted a periscope and alerted the destroyer *Ward*. The initial sighting was made about four hours before the air attack on the fleet, and the *Ward* sank a different miniature submarine about an hour before the general attack. This information was not processed in time for an alert to be issued.

Certain aspects of the battle that day shocked the United States. How could Japan have sent a major task force across 4,000 miles of ocean without being detected? Why were early warnings of the attack ignored (especially radar interception of the first attack wave)? How had the Japanese effectively used torpedoes in the attack when the US Navy had proved to its own satisfaction that Pearl Harbor was too shallow for airplane-launched torpedoes?

The Japanese themselves were astonished. They launched two attack waves with virtually no resistance from the more than three hundred US planes on the island. General Walter Short, the army commander charged with the air defense of Pearl Harbor, had ordered his planes parked in the middle of the runways at Hickam and Wheeler air bases, where they were destroyed on the ground by the first Japanese attack wave. Had the Japanese launched a third attack wave, or had they followed up with a landing on Oahu, they quite likely would have taken the United States out of the war. The Japanese were not prepared for success.

As it was, the United States suffered a terrible defeat at Pearl Harbor, much worse than the American public was led to believe. The entire fleet of battleships, the first line of the navy's defenses, were disabled or destroyed. There was effectively no Army air force left in Hawai'i. Military historians have pointed out that it was a great victory for Japan but not the decisive victory it could have been. The US carriers, which, rather than battleships, proved to be the major naval weapon of the war, were not at Pearl Harbor during the battle.

The importance of Pearl Harbor as a battle is that it transformed the United States from an isolationist state to the major world power it is today. We occasionally hear the sentiment that "the United States cannot be the policeman for the world," but in effect it is—and has been since December 7, 1941.

THE BATTLE OF ANTIETAM

US Civil War battlefields are threatened by the one landscape feature that geographers find especially appalling but that Americans are largely indifferent to: urban sprawl. Most battles in the 1860s were fought in rural areas, but the unchecked growth of suburbanization means that some battlefields like Manassas/Bull Run have been virtually surrounded by housing developments and shopping malls. **Answer 36d:** Even the formerly isolated battlefield near Sharpsburg, western Maryland, where the Battle of Antietam was fought, is threatened with urban sprawl.

The US Civil War was one of the bloodiest conflicts ever fought; more soldiers were killed in that war than in all other US wars combined. Antietam was the deadliest single day of the Civil War, with total casualties reaching twenty-three thousand. This was the first major battle of the war, and the casualties mounted because the infantry tactics employed had been made obsolete by the weapons used. The muzzle-loaded musket had long been the infantry's chief weapon. Muskets could be loaded and fired relatively quickly (two to four times a minute) but they were not very accurate weapons. Troops needed to be within a relatively short distance of their targets for the muskets to be effective. The 1861 Springfield rifled musket was the most common weapon used in the war, and it was much more accurate than its predecessors. The projectile used was the famed minié ball, which was designed to expand to fit the rifling in the barrel when the weapon was fired. This not only provided greater accuracy but the ball itself was truly a lethal projectile. Aside from being huge by modern standards, it also tended to expand on hitting a target, thus producing a savage wound. Amputations of limbs were very common in treating the wounded. Battlefield tactics would change later in the war, but at Antietam, the slaughter was horrendous.

Antietam should have been a decisive Union victory and should have destroyed Lee's army of northern Virginia, quite likely ending the Civil War soon after it had started. Union forces greatly outnumbered Lee's troops, and Lee had divided his army to take other objectives. Almost unbelievably, Union troops recovered dispatches from Lee that had been accidentally discarded. The Union knew that Lee's army had been divided! Moreover, this was Lee's first penetration into northern territory, and he expected support from Maryland civilians who hadn't been strong supporters of President

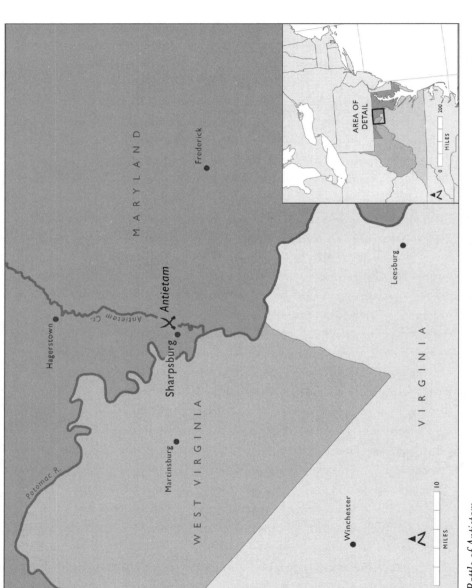

The Battle of Antietam

Lincoln. This proved not to be the case. General George McClellan, commander of the Army of the Potomac, failed to take prompt action when he caught Lee in a disadvantaged position and failed to use his reserve forces. The result was that Lee was able to achieve a stalemate and withdraw his army safely.

Antietam proved to be of great importance because of what it led to and what it prevented. Just five days after the battle, Lincoln issued a first draft of the Emancipation Proclamation. The Civil War was no longer just about preserving the Union but also about abolishing slavery. It was the beginning of an enormous change in the United States that is still under way today in which Americans of African ancestry would share in the same liberties fought for and achieved in the American Revolution and its aftermath. Antietam also thwarted the efforts of those in England and France who urged support for the South based, in major part, on the need for cotton to keep their textile plants running. It was clear that the South wasn't going to be unchallenged as an independent country, as European mill owners had hoped.

BUILDINGS

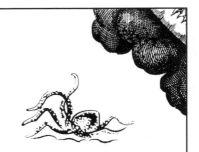

The geographer Carl Sauer distinguished between the natural and the cultural, or man-made landscape. It would seem to be easy to tell them apart, but some natural features (like **DRUMLINS AND ESKERS**) seemed to some to be so regular in appearance that they had to be man-made, while ancient man-made features such as mounds in Ireland and Wales as well as tells in the Middle East could easily be mistaken for natural features. Buildings are clearly man-made and important features of the cultural landscape.

Dr. G says: Drumlins and eskers are depositional features
of glaciers.

Since some of the earliest things built by man probably were related to religious practices, it's hardly surprising that some of the
most famous buildings in the world have religious significance.
Great cathedrals, such as St. Peter's in Rome, or famous mosques,
like the Sultan Ahmed Mosque (Blue Mosque) in Istanbul, are
examples of buildings constructed by the universal religions. Buddhism, too, has its important buildings.

Buddhism, the oldest of the universal religions (i.e., those that
seek converts), is based on the Four Noble Truths preached by its
founder. One branch of Buddhism that developed and flourished
in inner Asia was Lamaism. It became the principal religion of
Mongolia; at one time in the 1920s it was estimated that one-third
of all Mongolian men were Buddhist monks. It's not Mongolia,
however, with which we usually associate Lamaism, but rather Tibet.
Answer 37a: Until he fled Tibet in 1959, the Dalai Lama lived in
Lhasa, Tibet, in his palace known as the Potala. Today the Potala is
a museum.

The Mormons (discussed in an earlier chapter) believed that temple
worship would become an important part of their religious practice.
Worldwide, there are currently 134 Mormon temples and an additional 26 either under construction or planned. The temple built
in Temple Square in Salt Lake City is an impressive symbol of their
faith, but it wasn't the first temple they built. **Answer 37b:** The first
Mormon temple, and the only one built while the faith's founder,
Joseph Smith, was alive, is in Kirtland, Ohio. It was completed in
1836.

Not all famous religious buildings are large. **Answer 37c:** The most
sacred building in Islam is the Kaaba, a small, roughly cubical

building in Mecca, Saudi Arabia. When Muslims pray five times a day, they prostrate themselves toward the Kaaba. Muslims believe that the Kaaba was reconstructed by Abraham and his son Ishmael. The hajj pilgrimage, which each Muslim is required to make, if possible, requires that the Kaaba be circled seven times in a counterclockwise direction.

Royal palaces are among the most spectacular of buildings. With the decline of kingdoms and empires, surviving palaces usually become public buildings, most often museums that store and display art treasures. The Louvre in Paris, for example, has become the world's largest museum housed in a single structure. The Hermitage in St. Petersburg is perhaps even more spectacular and is remarkable for having withstood the horrible siege of Leningrad during World War II.

The Ottoman Empire, after its conquest of Byzantium (Constantinople) in the fifteenth century, continued in existence until 1921. The Ottoman Sultan ordered a palace built that at one time housed over three thousand people. **Answer 37d:** Today the Topkapi is a museum rather than a royal palace and overlooks the Bosphorus Strait and Sea of Marmara in Istanbul. Along with thousands of treasures from the Ottoman Empire, the Topkapi also holds relics sacred to the Islamic faith, including the sword and cloak of the Prophet Mohammed.

The Mughal Empire was the principal ruling force in the Indian subcontinent for about three hundred years, until it was officially displaced by the British Empire in 1858. The Mughals, like the Turks, originated in central Asia and were Islamic. They had been strongly influenced by the Persians, however, and brought that influence to India. The most famous building constructed by the Mughals is the Taj Mahal, in Agra, India. It's a mausoleum, built for the third wife of a Mughal emperor. The seat of Mughal power, however, was in Delhi, India, where the Mughals built the Red Fort. In the nineteenth century it also became the symbol of British power in India. **Answer 37e:** Today the Red Fort, while not quite a museum, is the leading tourist attraction in Delhi.

Cosimo de Medici built his palace between 1560 and 1591, and by 1765 it became one of the world's earliest museums housing treasures of the Italian Renaissance. **Answer 37f:** The Uffizi, in Florence, Italy, is one of the most famous and popular of all art museums. Unfortunately, its popularity exceeds its capacity—during the summers, crowds might have to wait five hours or more to enter. Since it contains works by every important Italian artist, many gladly endure the wait. Still, a common tourist lament is, "We went to Florence but couldn't get in the Uffizi."

At the time it was built to house the US Department of Defense, claims were made that the Pentagon was the largest building in the world. However, it no longer qualifies as such. **Answer 37g:** The largest building by volume in the world—nearly twice as large as its nearest competitor—is the Boeing Everett Factory building in Everett, Washington, which was originally built for assembly of the 747 aircraft. The Association of American Geographers sponsored a field trip to that building during its annual meeting held in Seattle. Among a host of other things, participants learned that parts for the 747 were manufactured in more than three hundred locations around the world, a vivid example of what the term *global economy* means. **Answer 37g:** The largest building by floor space is also relatively new: the airport at Dubai in the United Arab Emirates. Dubai is one of the most interesting phenomena anywhere. Aside from its airport and spectacular urban landscape, fewer than one-fifth of the population of Dubai is a native of any of the emirates. The largest single national group is Indian, constituting more than 40 percent of the population.

Certain buildings that are major tourist attractions (like the Uffizi) can permit only a limited number of entrants. Some, like the Eiffel Tower, are limited by the capacity of their elevators. **Answer 37h:** The most visited building in the world, according to several sources, will probably come as a surprise. It's the Lotus Temple in New Delhi, India. The temple is of the Baha'i faith, a religion originating in Iran (Persia) in the nineteenth century. It easily outdraws the nearby Red Fort and Taj Mahal. I have warned earlier that trivia questions that ask about the "most" or the "largest" should always elicit skepticism, and so it is with this category. Measuring the number of visitors to a building is hard enough when admission is collected and almost impossible when it's not. Recently the chief operating officer of the MGM Grand Hotel in Las Vegas (alleged to be the second largest hotel in the world) claimed that his facility (including the casino, restaurants, shops, and exhibits) received fifty thousand visitors per day. Shortly after their openings, both Penn Station and Grand Central Station in New York City claimed fifty thousand visitors daily. This is the same number claimed for the Lotus Temple! If you answered any of the large Las Vegas hotels, take credit for a correct answer! Sadly, the United States has turned its back on rail travel, so train stations are not visited often anymore, and the New York stations are no longer a correct answer.

It should be somewhat easier to measure size when we consider stadiums. The Greeks and Romans added a new kind of structure to the built environment when they introduced athletic events as public spectacles. The Colosseum in Rome survives from antiquity and even in disrepair is still a major tourist attraction. It was huge for its day, almost certainly the largest building in the world two millennia ago. By today's standards, it would be midsized since it seated about fifty thousand. The sporting events featured there would probably not be money makers today (with the lions eating up the prophets as they did). **Answer 37i:** The largest stadiums today are generally devoted to football or soccer. The largest of all is found where you'd least expect it: in Pyongyang, North Korea. May Day Stadium there has a capacity of 150,000. There's a much larger stadium, but it's now closed. Strahov Stadium in Prague, the Czech Republic, has been closed for about twenty years, but it has a capacity of about 220,000!

MOUNTAIN PEAKS

....

Question 38a: What is the second highest mountain in the world?

Question 38b: What is the highest mountain in Canada?

Question 38c: In what country is the highest mountain peak in the Western Hemisphere?

Question 38d: In what country is Mount Kilimanjaro?

Question 38e: What is the most dangerous high peak to climb?

Question 38f: What volcano erupted from a fissure in a cornfield and grew over a thousand feet high in a year?

....

Mountains are of central concern to both physical and cultural geographers. Mountain building and erosion of mountain slopes are among the most fascinating topics in the physical geography curriculum, while mountains as barriers to human movement and as home to many different peoples are the focus of interest to those who specialize in human geography. Mountain peaks, however, are probably of only peripheral interest to the science of geography. This is unfortunate because mankind has long been fascinated and awed by high places. Ancient religions looked to mountain peaks as the abode of the gods, or at least as places of spirituality. The Ten

Commandments and Zion are associated with mountains, and the Potala (mentioned in an earlier chapter) is at twelve thousand feet elevation. Peaks also challenge people (not just modern climbers). "Pike's Peak or Bust" was a famous slogan of westward movement in the United States, and Quechua (Incan) remnants have been found at high elevations in the Andes.

During my lifetime the most famous mountain ascent was made by Edmund Hillary when he was the first (as far as we know) to climb the highest mountain in the world, Mount Everest in Nepal, in May 1953. (It's unfortunate that Hillary was a New Zealander from Auckland, since, as we learned earlier, it's unlikely that most college students have any idea where New Zealand is.) Hillary was accompanied by Tenzing Norgay, a Sherpa and a native of Nepal, but for reasons that I don't understand, Hillary was named by *Time* magazine as one of the one hundred most influential people of the twentieth century but Norgay was not.

Everest was not an easy mountain to climb, but at least it was easy to find. Strange as it may seem, not all truly high peaks are in easily accessible locations. **Answer 38a:** This is true of the second highest mountain in the world, K2, at 28,200 feet. It's located on the border between Pakistan and China. The name alone is a hint of its inaccessibility. When peaks were listed on European maps of Asia, they were identified by a code, this peak being the second in a chain of peaks in the Karakoram Range, hence "K2." The idea was that later the local name for the peak would be used, once it was known. It turned out, however, that there apparently was no local name because the peak wasn't visible from any human settlement. The first expedition that attempted to climb it required a two-week trek just to get to the base of the mountain. K2 has been nicknamed the "Savage Mountain" not only because it proved very difficult to climb but because there's a high fatality rate among those that have attempted to climb it. The mountain was first climbed by two people you've probably never heard of (unless you're a climber yourself), Lino Lacedelli and Achille Compagnoni. Fortunately, they're from Italy, which my college students can locate because it looks like a boot (although some confuse it with Louisiana, which also looks like a boot).

Answer 38b: Quiz programs and trivia games in general love to ask about another mountain that's not easy to find: Mount Logan. At 19,500 feet it's the second tallest mountain in North America, after Mount McKinley (Denali), and the tallest mountain in Canada. Because Mount Logan is in Canada it's not surprising that many Americans don't know about it, but it's also surprising that some Canadians put it in the wrong place as well. It's located not in British Columbia but in the Yukon Territory. As mountains go, it's not a truly difficult climb, but its base is a bit off the beaten trail. When it was first climbed in 1925, the climbing team took sixty-five days to reach the base of the mountain from the nearest town! Mount Logan is still growing in altitude because of tectonic plate uplift. This may be true of other mountains as well, but the phenomenon has been observed best on this peak.

Answer 38c: The highest peak in the Western Hemisphere is Mount Aconcagua at 22,800 feet. On a good day it's easily visible from Santiago, Chile, but the peak is actually fifteen miles across the border near Mendoza, Argentina. The Andes chain, of which Aconcagua is a part, essentially walled Chile off from the rest of South America. It was only in the latter half of the nineteenth century that the country expanded to its current territory. Chileans sometimes refer to their country as a *pais de rincones*, a country of nooks and crannies. The mountain wall reinforces this idea to some extent, but the length of the country (the longest north-south distance of any country) ensures almost every kind of climate and rainfall regime, each tucked in its own little corner of the country.

Kilimanjaro is one of the world's most famous mountains, in part because it's snow-covered yet is located only about three degrees south of the equator. It's a huge volcano, with separate peaks that

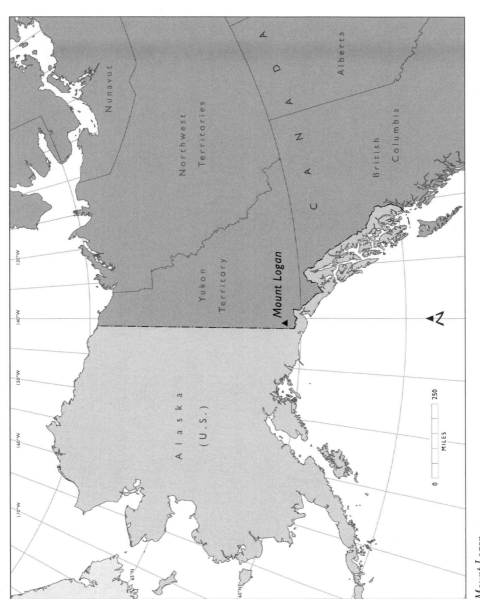

Mount Logan

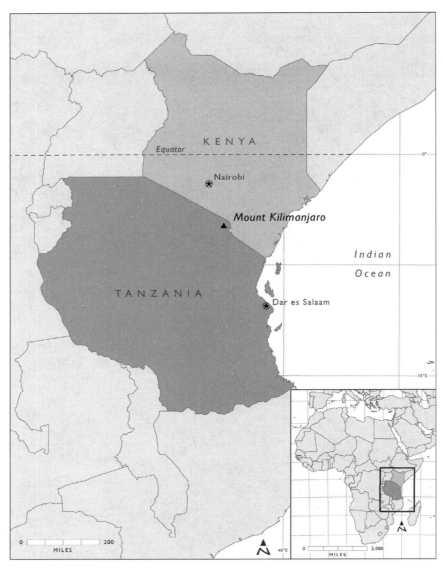

Mount Kilimanjaro

are either extinct or dormant. Its rise from the plain of East Africa is one of the most impressive sights on earth. I was once invited to a presentation on Kilimanjaro by a local geography society (which seemed to be more of a front for a tourist agency than any sort of "society"). Unfortunately the speaker had the mountain's location wrong and several times mentioned Kenya as Kilimanjaro's home. **Answer 38d:** After the presentation I mentioned to him that Kilimanjaro is actually in Tanzania and he replied that "most people prefer that it be in Kenya." This is an interesting concept, a sort of "let's have a democratic vote on where geographic features should be as opposed to where they actually are." Some of my students would have loved this concept!

Kilimanjaro is an interesting barometer of world climate. The surface of the mountain that is covered by snow and ice has declined significantly in recent years, and at least one prediction is that the famous "snows of Kilimanjaro" will be entirely gone by the year 2020.

Some mountains, like Kilimanjaro, are relatively easy to climb, but that doesn't mean that climbing them is safe. Even when specialized equipment isn't needed for a climb, any number of accidents can occur. Inexperienced climbers, moreover, often fail to accustom themselves to high altitudes and suffer from altitude sickness. **Answer 38e:** Of the high peaks in the world, Annapurna I, at about 26,600 feet, may be the most dangerous. The Annapurna Massif is, like Mount Everest, in Nepal, but for every one hundred successful ascents of Annapurna I there have been thirty-eight fatalities.

Volcanoes take different forms and behave quite differently in different parts of the world. The United States has, in the recent past, witnessed the explosive eruption of Mount St. Helens in Washington State, while thousands of visitors have flocked to the continuing eruption of Kilauea on the island of Hawai'i. Volcanoes in Iceland and Chile have recently interfered with air traffic as ash has been driven into the upper atmosphere. **Answer 38f:** One of the most interesting volcanoes in modern times, however, was Paricutin in Mexico. In 1943, an eruption began in a fissure in a cornfield in Mexico. Within a year, the resulting cinder cone had grown more than a thousand feet. Paricutin is an example, albeit a spectacular one, of the kind of volcanic activity that happens once and then is finished.

LAKES

Question 39a: *Are there really ten thousand lakes in Minnesota?*

Question 39b: *What is the name of the lake formed by the construction of the Aswan Dam on the Nile River?*

Question 39c: *Which Great Lake is entirely within the United States?*

Question 39d: *What is the world's deepest lake?*

Question 39e: *Where does the Bolivian navy carry out its training?*

Question 39f: *Where is the noted abolitionist John Brown buried?*

Question 39g: *The northward-flowing Jordan River flows from which freshwater lake to what saltwater lake?*

"Nature abhors a vacuum" so the old saying goes, but nature doesn't think that highly of lakes, either. In general, nature will work to destroy a lake so that when a map shows lots of lakes, it's a good bet that something has happened there relatively recently, before nature has had a chance to get rid of the lakes. So it is with much of Canada and northern Europe where glaciers came and went ten thousand to twelve thousand years ago. In an earlier chapter, the Winter War between Finland and Russia was mentioned. The number of lakes and former lakes (swamps) in eastern Finland is nothing short of astonishing—which is why the Russians waited until everything was frozen before they invaded. Much of Minnesota was under glaciation and the landscape is therefore also frequently punctuated with lakes. Are there really ten thousand lakes in Minnesota? **Answer 39a:** Admittedly it depends on how large a pond of water has to be to be counted as a "lake," but a reasonable measure (ten acres or more for a lake) yields a count of more than eleven thousand lakes in the state. "The land of eleven thousand lakes" just doesn't have the same ring! (This is, of course, the easiest question in this book, since it's a True-False question.) Minnesota also has a portion of the shoreline of the largest freshwater lake in the world (as measured by the surface it covers), Lake Superior.

Some lakes are artificial. The largest artificial lakes result from dam construction projects. The biggest and most famous of these in the United States is Lake Mead, the lake that formed behind the construction of Hoover Dam on the Colorado River. **Answer 39b:** There is also a huge lake behind the Aswan Dam on the Nile River. The portion in Egypt is called Lake Nasser, and the portion in the Sudan is called Lake Nubia. Since most lakes, including artificial lakes, have streams flowing in and out, the presence of a lake slows the speed of the water in the stream that created it. As the water slows, it drops the material it has been carrying and deposits it on the bottom of the lake, thus filling in (and eventually destroying) the lake. Since people who build dams don't want their created lakes destroyed, they're forced to constantly dredge the lakes to remove the depositional material.

The United States went through a huge dam-building era during the Great Depression. Dams were viewed as marvels of progress: they prevented floods and produced electricity. There are, however, costs associated with dams that either weren't understood when they were built or were ignored. We now may be entering a dam destruction era, as at least three dams in the Pacific Northwest have been removed in recent years.

While the decline of geography in public schools has produced much of interest (and disgust!), one of the most fascinating is the ignorance surrounding the Great Lakes. Older people, particularly from the Midwest, assume that when they mention the Great Lakes, others will understand what they mean. Among the exceptions to this was one of my freshman students who thought the Great Lakes were a musical group (and they may well be!) and another student, specializing in environmental studies, who thought it was unfair to call some lakes "great" and thus, as she put it, "to dis other lakes." There was also some confusion in the political realm recently when a politician tried to get Lake Champlain added to the list of Great Lakes. Despite the confusion, the Great Lakes stand out as an enormous resource shared between the United States and Canada. **Answer 39c:** And shared they are, all except Lake Michigan, which is the only Great Lake entirely within the United States. Lake Superior may be the biggest, but if Michigan and Huron were considered as one lake (as they probably should be), the resulting lake would be even larger than Superior.

Since nature tries to destroy lakes, it's often believed (and even proclaimed in some geography texts!) that there is no such thing as a really old lake. **Answer 39d:** While "old" is a relative concept (an important tenet of the AARP), Lake Baikal in Russia is old enough to challenge the idea of "no old lakes." It's at least thirty million years old and in several ways is one of the most incredible geographic features on earth. It contains the most freshwater of any lake—about one-fifth of the unfrozen freshwater on earth. It's also the deepest lake, with a maximum depth of over a mile! The water in the lake is noted for its clarity. In fact, it tends to be the standard against which the clarity of other lakes is compared. Unfortunately, the Russians decided to build a paper mill near Lake Baikal. The pulp and paper industry, left to its own devices, is close to the ideal pollution generating agency; my experience in living near two such plants is that they pollute ground water, surface water, and the air. Lake Baikal is indeed threatened.

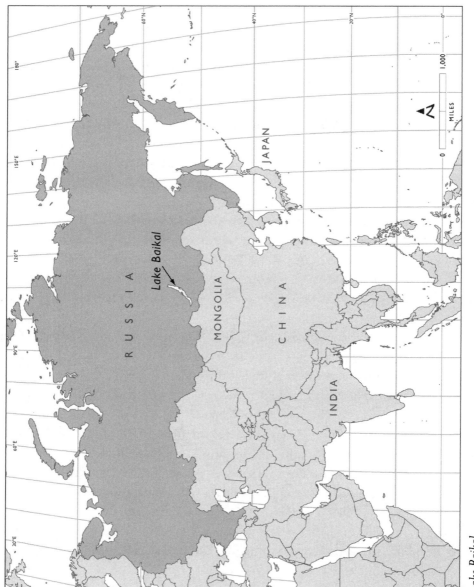

Lake Baikal

Lake Titicaca

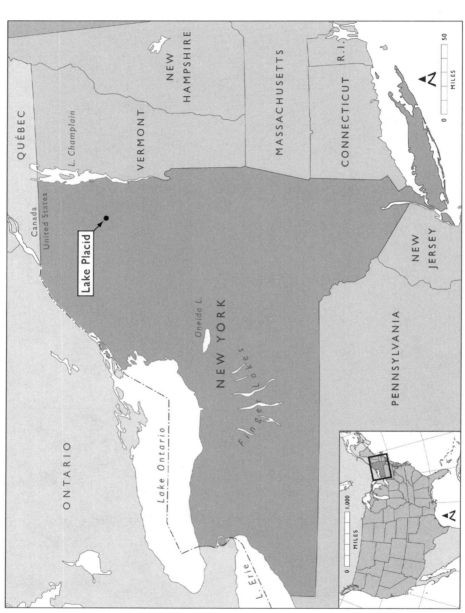

Lake Placid

Another reasonably old lake is shrinking and has dropped nearly two feet in mean depth over the past few years. Lake Titicaca, on the border between Peru and Bolivia, is one of the highest lakes in the world, at about 12,500 feet above sea level. The lake's level has been declining for some time and has now reached its lowest level since 1949. Core samples taken from the lake's bottom indicate that the lake has risen during periods of continental glaciation and dropped in volume during interglacial periods. **Answer 39e:** Landlocked Bolivia has a navy and conducts its naval exercises on Lake Titicaca!

There are few villages of 2,500 people that have attracted so much attention for so long as Lake Placid, New York. Only two places in the United States have twice hosted the Olympic Games: Los Angeles and little Lake Placid. The Winter Games of 1930 and 1980 were held in Lake Placid and, if that were not sufficient fame, perhaps the most famous incident in US Olympic history, the so-called Miracle on Ice, when the United States defeated the Soviets for the gold medal in hockey, occurred in Lake Placid during the 1980 games. As it turns out, Lake Placid should be well known even if it had never hosted an Olympics. Before the Civil War, abolitionist John Brown bought a farm near Lake Placid and attempted to start a colony of freed slaves. **Answer 39f:** When Brown was executed for treason at Harper's Ferry (then Virginia), he requested burial at his Lake Placid farm. So, when the refrain about "John Brown's body lies a'molding in the grave" is sung, recall that the site is Lake Placid!

One of the classic geographic trivia questions involves the Jordan River and two lakes that are a source and a destination. The Jordan River of biblical fame flows from the Sea of Galilee to the Dead Sea. Neither body of water is a true sea, but the Sea of Galilee is freshwater while the Dead Sea is salt water. *That* Jordan River, however, flows southward. **Answer 39g:** In Utah, the Jordan River flows northward from freshwater Lake Utah to the salty Great Salt Lake!

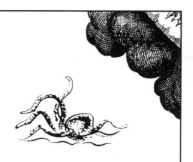

OUTLIERS

....

Question 40a: What is the highest mountain in Spain?

Question 40b: What countries own Greenland, Easter Island, and the Galápagos Islands?

Question 40c: What is the only Canadian province that is entirely an island?

Question 40d: In what state is Ellis Island, the famous immigration port?

Question 40e: What Hawaiian island is privately owned, and whose residents speak Hawaiian as their primary language?

Question 40f: Which is the only US state that borders only one other state, and what is the single state that it borders?

Question 40g: What two South American countries do not border Brazil?

....

By now you should be good (or at least better) at answering geographic trivia questions! This chapter will challenge you a bit more since it deals with outliers, things that buck the trend and are often unexpected. Unlike the easy True-False question about Minnesota, some of these questions require multiple answers.

Science, particularly science that deals with humans (as geography does) looks for trends, or more formally, measures of central tendency and associations. There is, for example, a strong association or relationship between energy used per person and income per person throughout the world. Some cases, however, don't fit the relationship. One term for these is *errors*, but scientists (who are often quite sensitive about such things) prefer to call them "outliers." In the example of energy use and income, some countries like those in the Persian Gulf or a Pacific island like Nauru are outliers: all these countries have relatively high income but little or no industrialization so they use little energy. Instead, they sell natural resources like petroleum or phosphates and have high incomes (at least as long as the resources last). Outliers are useful in helping us understand the flaws in our thinking about generalizations and trends. There are, of course, geographic outliers as well—places that may literally be outside the general ebb and flow of things or otherwise are exceptions to general rules.

Answer 40a: Spain has some beautiful and spectacular mountains, but the highest point in Spain is a true outlier: Mount Teide, which is more than six hundred miles from the Iberian Peninsula! Teide is the huge volcano that created the island of Tenerife in the Canary Islands (mentioned in an earlier chapter). Teide is about 12,200 feet above sea level and, since the Canary Islands are indeed part of Spain, it is by far the highest Spanish mountain! Tenerife is a surprise in other ways as well. It's very densely populated and has over 900,000 inhabitants, with an additional 5 million visitors annually. Tenerife is also notorious for another outlier: the worst aviation accident in history, when two 747s collided on the runway, killing 538 people. The details of the accident are mind-boggling since neither aircraft expected to be at Tenerife (a bomb-scare on Grand Canary Island caused a diversion of air traffic), neither plane could see each other, the ground controller could see neither plane, and there was no ground radar.

The era of colonialism is over, and while most former colonies have gained their independence, a few have opted for a status as dependencies or have become integral parts of the former colonizing power. Several of these territories are indeed outliers in the sense that they're at a considerable distance from the home country. **Answer 40b:** The largest example is Greenland, which became a Danish colony in 1814. Although it has obtained more self-rule in recent years, one fact helps explain why it's still a part of Denmark: its fewer than 60,000 inhabitants receive support from Denmark at a rate of more than $11,000 per person per year!

Answer 40b: Easter Island, or Rapa Nui, is owned by Chile, although it's more than 2000 miles from the nearest point of the Chilean mainland. Several publications claim that Easter Island is the most isolated place on earth, but don't bet your trivia stash on that possibility! Hawai'i, an outlier in its own right, is about 2,500 miles from North America. **Answer 40b:** Also partially in the South Pacific (the islands straddle the equator) are the Galápagos Islands, famous for their wildlife and for the research done there by Charles Darwin. Their location is a clue to their owner: Ecuador, which is about 500 miles to the east.

Due to its enormous size and its population clustered along the US border, Canada is bound to have a few outliers, especially in the north. Increasingly, however, the Maritime Provinces have become somewhat of an outlier. With the decline of the fishing industry in the Atlantic and the real threat that the secession of Quebec would isolate them from the rest of Canada, the Maritimes have the potential to become a true outlier. Of the areas on the Atlantic seaboard, Newfoundland is probably the most significant Canadian outlier. It's an island, of course, and only part of Canada since 1949—but it's not a province, only part of one. **Answer 40c:** The true island province of Canada is the home of Anne of Green Gables: Prince Edward Island.

About 100 million Americans have at least one ancestor or living relative who has passed through Ellis Island, in the Hudson River in New York harbor. Prior to the 1920s, immigration from Europe was only partially restricted, and 13 million passed through the facility at Ellis Island, with the peak year being 1905, when over a million immigrants entered. It's a pervasive myth that all were admitted: about 2 percent were returned to their home countries because they lacked sufficient money, were diseased, or were known criminals. For years, New York and New Jersey each claimed ownership of Ellis Island, although the general public assumed it was part of New York. **Answer 40d:** The US Supreme Court decided that most of **ELLIS ISLAND** was in New Jersey, thus making Ellis Island a small-scale example of an outlier!

> Dr. G says: If you answered both New York and New Jersey to this question, take credit for a right answer. If you answered only New York, however, no partial credit is allowed. Since New Jersey has legal claim to the majority of the island, take credit for a New Jersey answer.

While millions of visitors go to Hawai'i every year, only a few get to visit the Forbidden Island, Ni'ihau. The Robinson family has owned Ni'ihau, a few miles west of the island of Kaua'i, for many years and, at times, has been reluctant to permit even the governor of Hawai'i to enter the island. **Answer 40e:** Hawai'i itself is an outlier of the United States, and Ni'ihau is an outlier of Hawai'i where the Hawaiian language is the primary language spoken. In the years before World War II, the US Navy, fearing an airborne attack on Pearl Harbor, thought an attacking force might use Ni'ihau as a base of operations. Secretly, ditches were plowed across the island to prevent it from being used as a landing field. As luck would have it, on December 7, 1941, a Japanese fighter pilot crash-landed on Ni'ihau, was captured, escaped, and was killed a week later.

Two of the United States (Alaska and Hawai'i) are not bordered by any other states; they are true outliers. Forty states border at least two other states. **Answer 40f:** Only one state, Maine, borders only one other state; let's call it a semi-outlier! Almost half of my students correctly identify Maine. The tricky part is figuring out which state it borders. History majors may recall that Maine was originally

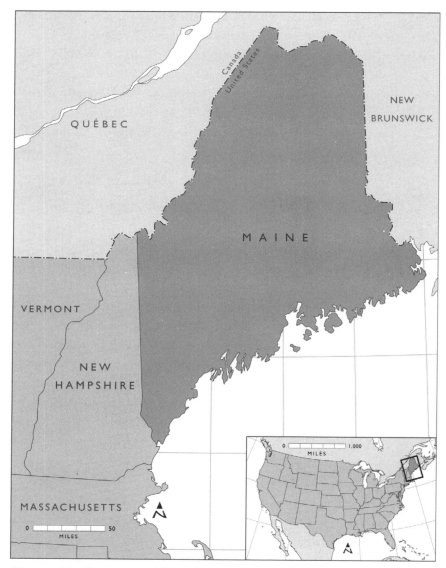

Maine and Its Connection to New Hampshire

South America Highlighting Outliers Chile and Ecuador

an outlier of Massachusetts but became a state to provide free-state/slave-state balance in the Missouri Compromise of 1820. Political science majors may recall the prediction of FDR's campaign manager, Jim Farley, of the 1936 presidential election when he quipped, "As Maine goes, so goes Vermont." (Which exactly predicted the election since only Maine and Vermont voted for the Republican candidate, with the others voting for FDR.) Neither Massachusetts nor Vermont, however, is the correct answer. **Answer 40f:** The single state bordering Maine is New Hampshire!

Brazil is, by area, the fifth largest country in the world, the largest in the Southern Hemisphere, and of course the largest in South America. Some economists consider Brazil to be the sleeping economic giant of the Western Hemisphere. Certainly its bold move to develop its interior by moving the capital from Rio de Janeiro to the new city of Brasilia attracted worldwide attention. The arrangement of countries on the South American continent makes it appear that Brazil is a huge central table, with the other countries set as chairs around it. **Answer 40g:** Two countries, however, are outliers and do not border Brazil: Chile and Ecuador.

CITIES

PLANNED AND FIAT

Question 41a: What is the youngest city in the world to reach a population of 1 million or more?

Question 41b: What two national capital cities were named for US presidents?

Question 41c: What island city, developed as a major port in the nineteenth century, became its own country?

Question 41d: What Florida retirement community has created a faux history of itself?

Cities are a central concern of geography, and urban geography has long been an important subfield. The processes that produce major world cities are often complex and subject to change as economic and political conditions change. In truth, it's difficult even to define cities, as their boundaries have become blurred and their functions altered. The largest migration in all of human history has moved a majority of the world's population from rural to urban areas. While urban geographers have done an excellent job in analyzing new roles of cities, new kinds of habitation forms, and new economies that have developed, they have often avoided stating the most profound truth about many urban centers today: they've become teeming pools of human misery.

We can generally explain the location of cities based on their roles as central places providing a range of goods and services or because they're nodes on important transportation routes. However, a few cities, many quite recent in history, are "fiat cities," which developed because governments ordered their creation. Sometimes these cities became important distributors of goods and services to regions, or transportation nodes, but these roles were subsequent to their creation. Urban geography, despite being a vital subfield of the discipline, has largely ignored this type of city, which is strange because fiat cities as a group have proven to be highly successful.

Planned cities and communities are similar to fiat cities but tend to involve careful consideration of the geographic aspects of location: **SITUATION AND SITE**. Retirement communities like SaddleBrooke near Tucson, Arizona, and suburban developments like Reston, Virginia (suburban Washington, DC), were built largely or entirely with private funds to generate a profit for the developers. Some retirement communities, like those outside Phoenix, are large and contiguous enough to qualify as cities in their own right. They contrast with Washington or Brasilia, which may have been carefully planned but were basically plunked down on the landscape to further a political policy.

> Dr. G says: *Situation* refers to the location of a place relative to other places, while *site* refers to local aspects of location (such as soils, bedrock, local climate, and topography).

Within the past few centuries a few large cities have sprung up largely because governments have decided they should be built. Saint Petersburg, Russia, was Peter the Great's "window to the West," an incredible city meant to tie Russia to Europe. Sometimes called the "Venice of the North" (along with several other cities), it's not sited where one would expect a great city. In fact its residents say the weather in St. Pe-

tersburg "is nine months of hoping and three months of disappoint-
ment." Russia's most dramatic fiat city isn't Saint Petersburg, however,
but Novosibirsk in Siberia. Novosibirsk was settled in the 1890s and
didn't become a city until after the beginning of the twentieth century. As the Soviets under Stalin moved industry and people to the
east in anticipation of Hitler's military invasion, Novosibirsk probably
became the Soviet Union's second largest city during World War II.
Answer 41a: By 1962, Novosibirsk passed a million in population, the
youngest city in the world to reach that level. Today it's Russia's third
largest city, behind Moscow and St. Petersburg.

The movement of the capital city of the United States from New York
to the specially constructed city of Washington, DC, was an early
example of the political compromise between Southern agrarian
and Northern commercial interest that was to characterize so much
of American history. Washington owes its entire existence to the federal government; none of the economic factors that shaped so many
urban locations operated to any important extent in Washington's
case. Brazil's capital, Brasilia, is another fiat city, but in this case its
location was not so much a compromise between regional political
interests as it was an attempt to develop Brazil's interior. A smaller
example is Australia's capital, Canberra, a twentieth-century city
built as a compromise between Melbourne and Sydney.

The slavery issue in the United States led to the creation of two
colonies in West Africa populated by freed American slaves. The first
was at Freetown (now in Sierra Leone) and the other a settlement
that came to be called Monrovia, after US president James Monroe,
who strongly supported it. Monrovia is partially a fiat city and partially what some urban geographers would call a "gateway city," built
to support trade between Africa and the United States and Europe.
When the country of Liberia was created, Monrovia became its
capital. **Answer 41b:** Washington and Monrovia are the two national
capitals named after US presidents.

Perhaps the most important nineteenth-century fiat city was built by the British on islands at the tip of their colony in Malaya. Unlike other fiat cities, the geographic situation of Singapore was superb. It commanded the shortest shipping route between the Indian Ocean and the Pacific. During World War II Singapore fell to a land-based Japanese force, thus inflicting on the British one of their worst and most **DISGRACEFUL DEFEATS** in their history. After the war, Malaya became the independent country of Malaysia. Malaysia has, since its independence, struggled with the issue of whether the country is a country of Malays or whether it's a plural society with the large number of Chinese and Indians sharing in the privileges of citizenship. **Answer 41c:** It was over this issue that Singapore, with its large Chinese population, **LEFT MALAYSIA** and became its own country in 1966. Since that time, Singapore has become one of the most successful countries in Asia and has developed one of the largest harbors in the world.

> Dr. G says: Winston Churchill was appalled at the surrender of tens of thousands of troops in Singapore although they outnumbered the Japanese invaders. The Japanese "secret weapon" was the bicycle, by which they were able to move quickly through jungle paths and repeatedly outflank the British on the Malay peninsula.

> Dr. G says: Sources often refer to this as the "secession of Singapore." It probably should be called the "eviction of Singapore."

Retirement communities have long existed in warmer climes of the United States and Europe (the "Prune Belt") but with the populations of both regions growing older, planned retirement communities have sprung up in many different locales, although they're still more heavily concentrated in states like Florida and Arizona. House types differ, but most communities have clubhouses with pools and gyms and offer a wide array of activities for their residents. Surrounding the communities are a range of services providing employment for the unretired, thus creating a "shadow population": where the elderly move, a migration of job seekers will follow. Restaurants offer foods familiar to the older residents and specialize in the senior penchant for poultry ("early bird special").

Most of the newer retirement communities are age segregated, with the strictest requirement being that no one under nineteen can live in them permanently. Some communities even have posses to patrol and ensure that young'uns don't sneak in. Of these communities, one of the largest and most unique is The Villages, located between Ocala and Orlando, Florida. Like other retirement communities, The Villages has plenty of golf courses and clubhouses—and now, probably upwards of 100,000 residents. **Answer 41d**: Uniquely, however, The Villages has created a faux history about itself and is designed somewhat like a theme park. It has, for example, an artificial lake and a lighthouse, both built recently, which are claimed to be over a century old. One of the most salient features about retirement communities (and one that should be of great interest to urban and political geographers) is the incredible loyalty that residents show toward the communities, faux history or not.

WHEVER
HAPPENED TO . . . ?

Question 42a: What is the modern name for Cambria?

Question 42b: What are the current names of former British Honduras and British Guiana?

Question 42c: What is Van Diemen's Land now called?

Question 42d: What happened to Afars and Isis?

Question 42e: The names of what two countries in Africa start with the letter "Z"?

Giving a name to a place not only endows that place with recognition but acknowledges that it's part of our life. Some places have symbolic meaning for whole nations; others have significance for perhaps only an individual or two. Place names change, and when they do, it usually means that a major upheaval has occurred. Perhaps a conquest has occurred, or perhaps a language has become extinct and the meaning of the place has been forgotten, or more likely in today's world, a developer has endowed a place with a name that gives it market appeal. Sometimes place names change for reasons that seem absurd. A building in which I had worked had its name changed because some people didn't like the research findings of the man for whom the building was originally named. In another case, a former professor of mine never forgave the state of Pennsylvania for changing the name of Mauch Chunk to Jim Thorpe. He had nothing against the great athlete, but Mauch Chunk, aside from being a colorful name, happened to be the name of a rock formation and he taught **GEOMORPHOLOGY**! Most of us have become used to places changing names, and we know that Iran used to be Persia, Thailand used to be Siam, and Ethiopia was once Abyssinia.

> Dr. G says: Geomorphology is the study of land forms. It's the science that explains why the earth is not a featureless plain—why it has "ups and downs," which we can call "relief."

The ancient names for the British Isles have endured and are used in a number of ways. "Albion," although perhaps of Latin origin, is an older name than the Roman name "Britannia" for England or perhaps for all of Great Britain. It was proposed at one time that the country we know as Canada (still south of Detroit!) be called New Albion. Hibernia (Ireland) and Caledonia (Scotland) live on in the names of social clubs or of other things named after them: New Caledonia, an **ISLAND** in the South Pacific, or *New Hibernia*, a publication that reviews Irish literature, are examples. Cambria, however, is lesser known, at least as a place. **Answer 42a:** "Cambria" is the ancient name for Wales and is of special interest to geographers and geologists. The Cambrian period is the first, or oldest, period in the Paleozoic era and is named after rocks identified in Wales.

> Dr. G says: Actually, New Caledonia is a collection of several islands.

Forget the geopolitical implications for the moment—the British Empire was a huge business enterprise. In the classic colonial model, the colony provided raw materials and markets and the mother country did the manufacturing and built the infrastructure in the colony to support the enterprise. It had to direct the building of the ports and the railroads and to educate and train the colonial population to run (if not govern) the enterprise. Once this was done, however, why continue to spend money to sustain the colony? The colony will still grow the coffee and bananas or provide the tin ore—what else can they do once all the infrastructure is in place and the plantations established? The disappearance of the colonial era (and especially the British Empire)—which is often explained by rising national aspirations in the colonies, the leadership of men such as Gandhi, or the sympathy of the British people toward the desire for independence— is perhaps better explained as a business decision: colonies were too expensive to maintain and, for some time at least, they would remain colonies even if they had a veneer of independence.

Some colonies maintained their names: Nigeria is still Nigeria, Cyprus is still Cyprus. Others, however, acquired new names. **Answer 42b:** British Guiana became the independent country of Guyana, not a major change. Guyana, however, will remain in American memories for some time as a result of an incident that happened there in 1978 when 918 Americans, including at least 270 children, were murdered or committed suicide at the direction of People's Temple leader Jim Jones. **Answer 42b:** British Honduras has become Belize, the northernmost country in Central America. Both Guyana and Belize retain English as their national languages.

In the seventeenth century, the Dutch were probably the world's greatest shipbuilders. The highly lucrative trade that they carried on with the East Indies (modern Indonesia) was facilitated by some of the largest sailing ships ever built. Some of these ships, lacking the ability to measure latitude (as discussed in an earlier chapter), failed to turn north soon enough and ended up shipwrecked on islands near Western Australia, or even on the Australian continent itself. This is not, however, what we mean when we say that the Dutch discovered Australia! The first landing on Australia by a European (with an intact ship) was by the Dutch in 1606, but it wasn't until 1642 that another Dutchman, Abel Tasman, on his way to discovering New Zealand, encountered a land that he named Van Diemen's Land. After Captain Cook landed in Botany Bay and put eastern Australia on the map, it was assumed that Van Diemen's Land was a southern promontory of the Australian continent. It wasn't until 1798–99 that Van Diemen's Land was circumnavigated and it was realized that it was an island, separated from the Australian continent by Bass Strait. **Answer 42c:** Today the island is an Australian state known as Tasmania.

Some years ago I was forced to confront a reality that faced me in the introductory geography class I taught. A significant number of students had very poorly developed mental maps of the world—so poorly developed, in fact, that they had no idea where much of anything was. When I would mention a country like France, for example, many students didn't know what countries were near France, what continent it was on, or how big or small it might be. (Of course some students had well developed mental maps of the world—some perhaps better than mine.) Recalling my own educational background, I realized that I knew where most countries were by fifth grade—and I wasn't going to teach fifth grade material in a college classroom! So I decided that all students would be responsible for knowing the location of any country mentioned in the textbook or in my lectures and that I would test this knowledge as an add-on to regular exams. An atlas was a required text, so they certainly could find the countries. Ninety-five percent of the students accepted this course requirement without complaint or comment (or dropped the class). I assured the class at the onset that I would only probe their knowledge about the location of prominent, important, or in-the-news countries and that I wouldn't ask about relatively obscure countries . . . like Afars and Issas.

Students have learned not to trust their professors, so it was widely assumed that the one map quiz item that would certainly be asked would be "Where is Afars and Issas?" The problem was that no one in the class could find such a place. Many assumed I'd made it up. **Answer 42d:** I didn't, but by the time I mentioned Afars and Issas in class, it had changed its name and had become the country of Djibouti! A gold medal was won in a recent Olympics Games by a Djibouti runner. According to press reports, it took almost three full days for the information to reach the country. So if you want to get away from it all, Djibouti may be the place!

The popular view of Africa as rainforest and jungle has been influenced by the reality of central Africa. This was one of the last areas to be colonized by Europeans. Incredibly, a vast area known as the Congo Free State was given to an individual, King Leopold II of Belgium. The country of Belgium itself had little interest in the Congo and Leopold took it on as a private fiefdom. At first, Leopold nearly went bankrupt with his Central African holding, but gradually it began to pay off in a big way, particularly as rubber became important in manufacturing and international trade. The on-the-ground reality was that the rubber and other Congo industries prospered on the back of a native population that was all but enslaved. Eventually, the Belgian government took over the area, improving things somewhat and changing its name to the Belgian Congo.

When the colony became independent in 1960 it may have been the least prepared of any colony for independence—put another way, the Belgians may have been the least interested colonizers in the welfare of the people it colonized. The independent country of Zaire has been beset by internal political strife for most of its existence. In the late 1990s, it changed its name to the Democratic Republic of the Congo. **Answer 42e:** So if you're looking for African countries beginning with the letter "Z," look to Zambia and Zimbabwe . . . Zaire is no more.

THE WILD WEST

Question 43a: What Illinois town, site of the westernmost battle of the American Revolution, is now virtually a ghost town?

Question 43b: What border dispute involved the cry "Fifty-four Forty or Fight!"?

Question 43c: Where did the "dead man's (poker) hand" originate?

Question 43d: What game, associated with the west, originated in Robstown, Texas?

Question 43e: Where is Geronimo buried?

Europeans who settled anywhere in the New World confronted something that was all but unknown in their homelands: a settlement frontier. Some of the earliest sketches and drawings of both North and South America portray a narrow band of beach, a few crude dwellings . . . and a wall of trees appearing as an impenetrable barrier. Literary works, particularly those in Latin America, describe homes as islands of civilization surrounded by a sea of foreboding, even evil, wilderness. The frontier was not so much something to be expanded or settled as it was to be tamed and conquered.

In the case of the colonial United States the boundaries of the settlement frontier were contested between the French and the British, with various Indian tribes allied with each country. Warfare and violence were associated with the west even when "west" meant areas of current western Massachusetts. This tradition of the frontier as a violent place followed the settlement frontier across the North American continent. Aspects of it, including issues of gun ownership, are still very much with us today.

In the Treaty of Paris that ended the French and Indian War in 1763, the British gained undisputed control of the area between the Mississippi River and the Appalachian Mountains. Almost immediately thereafter, the area was proclaimed off limits to settlement. The land, especially in the Ohio River Valley, was to remain Indian land. The French had earlier established settlements in this territory, the most important of which was Kaskaskia, on the banks of the Mississippi, and now (**HOWEVER PRECARIOUSLY**) in Illinois.

> Dr. G says: The Mississippi River has both flooded and changed course since the founding of Kaskaskia. The town, not in its original location, is actually now on the other side of the river.

Kaskaskia was an interesting town that in 1763 was far removed from the thirteen colonies on the Atlantic coast and considerably upstream from the French city of New Orleans. Its population was estimated at around seven thousand, an amazingly large frontier settlement that, despite its newly acquired status as British territory, remained largely French and Indian. During the American Revolution, George Rogers Clark led a force that seized Kaskaskia from the British and took British settlements in Indiana as well. **Answer 43a:** Today Kaskaskia has barely avoided becoming a ghost town. Its 2010 population of fourteen is a considerable percentage increase from its population of nine in 2000!

Fighting two or more wars at once is something to be avoided, as numerous national leaders have discovered. US president James Polk confronted this possibility during his term of office (1845–1849). In the southwest, Mexico refused to allow the United States to annex Texas. In the northwest there was a major border dispute with Britain over the border between the Oregon territory and what was to become the province of British Columbia. American sentiment seems to have focused on the northern dispute. **Answer 43b:** Polk had been elected on the theme of "Fifty-four Forty or Fight!," a reference to the line of latitude that many Americans hoped would divide the United States and British North America (now Canada). The possibility of a third war with Britain in a sixty-year period loomed large.

The term *Manifest Destiny*, which had first been used around 1811 to support American expansion into Ohio, was now used again. Its exact meaning is unclear, but basically it implied that anywhere in the Western Hemisphere where American interests were at stake, the Americans would go and stake a claim. Certainly in Polk's administration it implied not only westward expansion but expansion north and south as well.

Polk went to war with Mexico and brought almost all of the **PRESENT SOUTHWEST** into the United States. In the case of the northern border, however, Polk retreated from his campaign position and agreed on a border that would follow the forty-ninth degree of latitude, or five degrees and forty minutes south of the original claim. As we know from an earlier chapter, this is a straight line all the way from the (then) Oregon Territory to Minnesota and the Northwest Angle.

> Dr. G says: The portion that Polk didn't add to the United States in the Mexican War was the Gadsden Purchase, which later in this chapter we learn introduced the US Army to the Apaches.

The Wild West is famous for its stories and legends about heroes and villains. James Butler Hickok, better known as "Wild Bill Hickok," is an example. Hickok was born in Illinois, fought in the Civil War, and became a lawman in Kansas and Nebraska. He was also a scout and a gambler. **Answer 43c:** One of Hickok's greatest legacies revolves around his death. He was killed while playing poker at a saloon in Deadwood, in the Dakota Territory, now South Dakota. According to legend, his poker hand contained a pair of aces and a pair of eights, and these became memorialized as the "dead man's hand." The legend has been refined to the point where the eights and aces were in the two black suits. There is an ongoing controversy, however, about what the fifth card in his hand actually was—or if he lived long enough to have it dealt to him.

Gambling has long been associated with the Wild West—it seems that every Western movie ever filmed has at least one scene of gambling in a saloon. A place that was once a pleasant stopover on the Mormon Trail between Salt Lake City and California lost the reason for its existence in the early twentieth century when the spring that watered "the meadows" for which it was named dried up. Las Vegas, Nevada, nearly became a ghost town until gambling was legalized there. It has subsequently become the greatest gambling mecca ever created, far surpassing the fame of places like Monte Carlo. Las Vegas Boulevard, better known as "The Strip," has some of the largest and most spectacular hotels in the world. The choreographed fountains in front of the Bellagio Hotel are enough to amaze even the most jaded!

In the 1960s, a new gambling game was introduced to Las Vegas. It took some time to grow in popularity but it may now be the most popular form of poker and perhaps the most popular gambling game in the world. Most commonly called "Texas Hold 'em," there's some doubt as to where and when the game originated but certainly none in the collective mind of the Texas State Legislature. **Answer 43d:** It declared that the game originated in Robstown, Texas!

The Apache known as Geronimo is in both legend and fact one of the most important figures of the Wild West. The Apaches lived in several areas of the Southwestern United States and had waged an unrelenting war against Mexico as far back as the seventeenth century. The Apaches killed thousands of Mexicans and destroyed scores of settlements just in the early 1800s. Geronimo's wife, children, and mother were killed by Mexican troops. In his biography, Geronimo claimed that he didn't know how many Mexicans he had killed and that, as a general rule, he didn't think they were worthy of being counted.

In the agreement that was associated with the Gadsden Purchase, the United States agreed to defend the Mexican population against Apache raids. The United States thus took on the burden of warfare against the Apaches. Some sources cite Geronimo as a chief or a war chief, titles that he wouldn't have claimed for himself. He proved himself, however, to be one of the most dangerous and elusive enemies that the US Army had ever fought against.

His surrender of his small band has engendered controversy from the time it occurred to the present. Geronimo himself claimed that he was deceived by General Miles, who accepted his surrender. Geronimo was transported to Florida—along with members of the Apache Scouts, a unit of the US Army, who had fought against Geronimo and helped arrange his surrender. The Scouts and Geronimo were held as prisoners of war. In his latter years, Geronimo became a Christian and joined the Dutch Reformed Church. He was later expelled from the Church for gambling!

Answer 43e: Later Geronimo was moved from Florida to Alabama, and then to Fort Sill, Oklahoma. He was never allowed to return to the Southwest. He died at Fort Sill and was buried there. There are claims that Yale University's Skull and Bones Society acquired Geronimo's skull from Fort Sill, but there's no firm evidence to support these claims.

CHAPTER 44

SPATIAL DIFFUSION

....

Question 44a: What disease, believed to have started in the United States, killed more than 100 million worldwide during 1918–19?

Question 44b: Where was the first traffic light used?

Question 44c: In the modern era, what US state was the first to introduce a state lottery (1964)?

Question 44d: What cartoon character was banned in Finland?

....

We're not in the Stone Age (at least not yet), the Enlightenment, the Middle Ages, or the age of exploration and colonization. All these "ages" were labels assigned by historians well after the fact. What age are we in now? A good guess might be the age of change. Things are changing at an incredible pace all around us. When I began my teaching career, ditto machines, typewriters, and movie projectors were essential tools. Recently, I had to prowl about in a Dumpster searching for an irreplaceable documentary movie that had been discarded by a newly hired librarian. I managed to rescue it, but when I brought it to class, along with a movie projector I'd found in an old closet to my classroom, the students had no idea what the machine was or what it did!

As described in chapter 14, geography is in a princely position when it comes to describing and analyzing change. Even a series of maps is a basic tool for looking at change. In doing research for an earlier chapter on the American settlement frontier (chapter 43,"The Wild West") I was struck by the fact that historians writing about the frontier hardly ever used maps—and when they did they were simple locational maps with no attempt to use the maps to analyze or explain the pattern of change that is intrinsic to the very idea of a settlement frontier. When geographers deal with change, they often call it "spatial diffusion," the way in which things and ideas spread across a landscape.

Three general types of processes have been identified in spatial diffusion: contagious, hierarchical, and relocation. The first is probably the most common and the one we most readily think of when we think about change spreading. In *contagious* diffusion, things spread from one receptor to the next, perhaps not always smoothly and evenly but like ripples spreading across a pond (perhaps with a lily pad or a small island interrupting the ripples). *Hierarchical* diffusion implies that ideas or things flow through a hierarchy, often of cities, but sometimes the hierarchy can be people: administrators, government officials, businessmen. *Relocation* diffusion means that change occurs not because more receptors accept the change but because the receptors themselves move from one place to another.

Religions are sometimes used as highly generalized examples of these three types of spatial diffusion. Islam spread directly across the landscape, at least in its early days of expansion. It was carried by a conquering army, so contagious diffusion was hardly surprising! Christianity began in a hierarchical process, through the major cities of the Roman Empire. Judaism, since it doesn't seek converts, spreads because its adherents move—thus, relocation diffusion.

During World War I, a **PANDEMIC** of influenza developed that eventually killed more people than the black death that devastated Europe in the fourteenth century (see chapter 29, "Medical Geography"). It was an unusual disease in several ways. First, unlike most flu epidemics that are usually hardest on the very young and old, this variety of flu also attacked those in the prime of life. One estimate is that it killed between 8 and 10 percent of young adults worldwide in the 1918–19 period. Second, as it swept through communities, it produced devastatingly severe symptoms among those who contracted the illness first but caused different, and sometimes milder, sets of symptoms among those who came down with the flu later. Third, many public health officials, in their public statements, minimized the severity of the disease even in the face of mortality rates that ran 30 percent and higher among those who came down with the disease. Most Americans didn't understand that they were in the midst of the most severe public health threat in history. Those that suspected the worst were often dissuaded by newspaper reports.

> Dr. G says: A pandemic is a worldwide outbreak of a disease.

This flu, which was commonly called the Spanish flu, didn't originate in Spain. It probably started in the United States and was spread rapidly by the mobilization of troops for service in World War I. It was a classic case of contagious diffusion and spread at a rate that overwhelmed hospitals in the eastern United States. I once asked a large class of mine to ask their grandparents if they had any memory of family members who had died of flu, particularly in the peak months of September and October 1918. I was amazed at two results: (1) a substantial number of families, even among those who were born outside the United States, had members who died at that time, and (2) a number remarked that, yes, they had family members die from the flu at that time, but, no, they hadn't realized that so many others were affected.

Medical authorities at the time fooled themselves into believing (as they had with yellow fever) that they were dealing with a bacteria instead of a virus. This alone prevented the development of an effective vaccine at the time of the outbreak. While the contagion missed some places (Australia, for example, was only slightly affected) as happens with some spatial diffusion processes, it's difficult to explain why some places were very hard hit and others were unscathed. **Answer 44a:** The total number who died of the flu in 1918–19 has been estimated between 20 million and 100 million. The upper limit of 100 million is derived from the latest study and seems the most reliable. Some authorities have expressed the opinion that the disease was so severe and widespread that it had the potential to destroy human civilization. While this view could be considered alarmist, it helps to explain why public health officials have been so concerned about swine flu, bird flu, and H1N1 in recent decades.

The spatial diffusion of the use of traffic lights illustrates hierarchical diffusion. An invention that regulates traffic flow is naturally going to be used first where there is significant traffic flow. **Answer 44b:** Sure enough, the first recorded use of a traffic light was in London in 1868, when London was the world's largest city. The London case, however, involved only one signal. Although a traffic light system was developed in Salt Lake City around 1912, the first municipal use of a system was in Cleveland, Ohio, in 1914. This idea spread to major cities and eventually trickled down to small towns. Sometime in the early 1970s, a news story emanated from the island of Kaua'i, Hawai'i, reporting that they had received their first traffic light!

As ideas move through time and space, they're often modified to meet local conditions or accommodate changing circumstances. Sometimes, too, new ideas don't spread at all (at least not at first). In 1907, major league catcher (and future Hall of Famer) Roger Bresnahan was hit in the head by a pitched ball. Bresnahan, who had already invented shin guards for catchers, invented a batting helmet. As far as I can tell, he was the only one who wore one, and he was derided for it. In 1974, however, the Pittsburgh Pirates became the first Major League team to require batting helmets. Now they're used not only by batters but also by base runners and even by some fielders and base coaches. The helmet has been modified over time, and an ear flap has been added to the original design.

Generally, when new things or ideas are introduced and diffuse widely, there will be a small number of people who are "early acceptors" and others who will hold out as long as possible. My aunt was a classic early acceptor. She was the first person I knew to have an electric range and, in the 1950s, when an "ironing machine" was introduced, she bought one. She may have been the only one that did: ironing machines were even less successful than the Edsel and the New Coke. My mother, however, was a late acceptor. Her resistance to TV, automatic washers and dryers, and frostless refrigerators was nearly legendary.

Sometimes new things spread by a pattern described by geographer Larry Brown as "supply side" diffusion. The decision to start a diffusion process is made by a provider or a government agency attempting to bring about change. Promotional materials and other forms of marketing are often involved. State lotteries are a good example. **Answer 44c:** Although lotteries in one form or another existed in the nineteenth century in the United States, the first state to adopt a lottery in the modern era was New Hampshire, in 1964. Most other states followed suit and today only Alabama, Alaska, Hawai'i, Mississippi, Nevada, and Utah do not have lotteries.

Our discussion of spatial diffusion wouldn't be nearly complete unless we considered barriers to the spread of new ideas. All new things are born within a particular culture, and that culture stamps the new idea or thing with its own characteristics. Even broad changes in societies that we label with such terms as *industrialization* and *modernization* involve a range of things that were developed first in Western Europe, or in Euro-American culture. When these ideas reached other parts of the world, they've posed a real challenge. They were developed in particular cultures with certain institutions (say, a system of land ownership and control) and value systems (concerning, for example, who inherits land upon the death of an owner). The non-European culture receiving the new ideas may have entirely different institutions and value systems. The recipients of the new ideas say, "We want improvements in our standard of living . . . if nothing else, we'd like our children to survive infancy . . . but at the same time, we want to remain what we are (Africans, Asians, Pacific Islanders, etc.) without changing our culture." Most often, values and institutions in a receiving culture have to be changed to accommodate change—sometimes slowly, sometimes reluctantly. Some ideas, however, will be resisted strongly: even the traffic light, a real symbol of modernization, has been strongly resisted in some areas of the world.

Sometimes new things are resisted for unusual reasons. The names of newly introduced products may alienate consumers in a different culture. According to legend (at least), the Chevrolet Nova didn't sell well in Spanish-speaking countries since *no va* implies that the car doesn't run! **Answer 44d**: Finland, otherwise a bastion of progressive thought, banned publications that portrayed Donald Duck . . . on the grounds that he didn't wear pants!

PARKS

Question 45a: What was the first national park in the world?

Question 45b: What's the world's largest national park?

Question 45c: What municipal park featured Southdown and Dorset sheep grazing on its grass until they were removed in 1939?

Question 45d: At what park can you find around ninety thousand impalas at any given time?

Geographers study parks at a variety of scales. One geographer I know has designed a small municipal park, scattered with rare and beautiful plantings, a biogeographer's dream. Two others I know have studied the process of national park creation: one proposed park actually became a park but the other did not. Despite the size difference between a small municipal park and a national park, it's possible to see a feature common to both of them: they attempt to keep land from being developed by commercial interests, but they intend the land to be used by the citizenry in some fashion rather than simply "preserved." This doesn't always work out as intended.

Bumper-to-bumper traffic at peak times in some of our national parks raises the question of what is actually being preserved. Certainly a sense of serenity and enjoyment of nature is hard to achieve with overcrowding. Competition for space in some of our municipal parks has resulted in serious turf battles. This seems particularly true as more and more organized youth sports leagues vie for practice space. In my experience, the occasional "out-of-control" adult at a Little League game that makes the six o'clock news palls in comparison to the brawls that accompany conflicts over practice fields. In the community where I grew up, the parks we used to play in are now off-limits to most activities because of liability issues. These limitations are of little concern anyway because the automobile traffic that circles the fields makes access difficult and most activities (unless they involve exhaust fumes) impractical.

My favorite park exemplifies the general situation with parks. Adirondack State Park is the largest park in the conterminous United States and is actually larger than several states. The New York State Constitution has declared it "forever wild," and the state's voters have repeatedly denied even tiny attempts to violate that principle. Indeed, substantial areas in the park are wilderness; on the other hand, they've actually built an interstate highway through the middle of it! The management of the park, master of compromise, has been like that for decades: reintroducing wildlife like the moose and lynx but also authorizing power dams on rivers.

Perhaps the problem is that most parks are publicly owned and therefore subject to the vagaries of politics. The fear has always been that private interests would destroy the value of parks, but is that always the case? Parks seem to have been based originally on large tracts of land owned by aristocrats and used for their private enjoyment. Public parks are simply the democratization of that concept. The Royal Parks in London are truly spectacular, yet they're not really publicly owned. Green Park (47 acres), Regents Park (410 acres), St. James's Park (58 acres), and Hyde Park (350 acres) were originally the monarchy's hunting preserves. The fact that they merge and that London doesn't have high-rises means that you can walk long distances through them and forget you're in the midst of a large city.

Answer 45a: The first national park in the world is Yellowstone Park, mostly in Wyoming but spilling over a little into Montana and Idaho. Yellowstone is a supervolcano, similar to Kilimanjaro in Tanzania. Its location was a little too far south to have been visited by the Lewis and Clark expedition and a little too far north to be on the routes to Oregon or California. Reports of its existence were dismissed as myths or fantasies until after the US Civil War. Shortly after Yellowstone was officially "discovered," the US government, during the administration of President Grant, took possession of the area in 1872. Asserting this ownership required the intervention of the US Army. Creating the administration of Yellowstone and, eventually, a national park system, took many decades. The National Parks in the United States are truly one of the country's proudest accomplishments. Recently, political groups have called for the sale of the National Parks to pay off the national debt, and while privatization of the park land has been sought since the first parks were established at Yellowstone and Yosemite, it seems highly unlikely that the parks would fall into private hands now.

Since the creation of Yellowstone, the idea of national parks has spread worldwide. In fact, the final question of the 2011 National Geographic Bee concerned the name of the **NATIONAL PARK ON THE SOUTHERN SIDE OF MOUNT EVEREST IN NEPAL**; both finalists answered it correctly! The largest national park in the world, and one of the most recent (1974), is considerably larger than Yellowstone. **Answer 45b:** In fact, it's larger than more than 160 countries in the world! It's Northeast Greenland National Park, and aside from being immense it's also characterized by a complete absence of human population. It does however, have a substantial population of musk oxen, another

Dr. G says: If you must know, it's called Sagarmatha National Park.

(mostly) New World species that seems ideal for domestication (as far as I know, this hasn't been attempted; see chapter 23).

The man most associated with municipal parks in the United States is Frederick Law Olmsted, who along with his business partner, Calvert Vaux, designed parks throughout the country. Olmsted's most famous park is New York City's Central Park, covering over eight hundred acres in Manhattan. Vaux and Olmsted viewed the park as a means of making an idealized nature available to all—a sort of landscape democratization of New York. The many rules governing use of the park meant that Central Park was to be observed rather than used. The park has had its ups and downs over the years, sometimes flourishing, sometimes entering into periods of decay. Mayor LaGuardia appointed Robert Moses to rescue the park in the early 1930s, with excellent results. Moses made the park not only a place of tranquility in the midst of a huge metropolis but also a place of recreation with numerous sports fields and children's playgrounds.

After World War II New Yorkers became disenchanted with the amount of crime occurring in Central Park. One could learn from those who grew up in New York City how devastating the negative perception of the park was to them. Currently, however, the park is enjoying a new era, with crime down substantially both there and in other areas of Manhattan. Central Park is again a source of considerable municipal pride.

When you read the trivia question about British sheep varieties grazing in a park and the elimination of the sheep just as Britain entered World War II, you probably thought the park in question was one of London's Royal Parks. **Answer 45c:** The sheep, however, grazed in New York's Central Park from the 1860s until 1939!

Not all parks, of course, are public or royal. Amusement parks have attracted crowds for generations. Two of the most famous older parks are Tivoli Gardens in Copenhagen, Denmark, and Coney Island in New York City. Although both parks have similar attractions, Tivoli oozes sophistication and features fine dining as well as a Ferris wheel and kiddy rides, while Coney Island has a rather seedy ambiance with gargantuan rides. Parks featuring roller coasters have become common in the United States, and bus tours take families from one roller coaster park to the next.

Disney parks are in a class by themselves. Geography textbooks occasionally mention these parks, and in some of the books I've read, Disneylands are treated with derision and sarcasm. Strangely, the specific complaint most often brought against the Disney parks is that they create faux landscapes; this is precisely what most municipal parks do and is exactly what Vaux and Olmsted tried to do in their parks. This negative attitude is mirrored in the history of the Disneyland built in the suburbs of Paris. French government leaders and labor unions initially denounced the park (although this is the norm for everything from the Eiffel Tower to McDonald's) and at first the public seemed to boycott it. Now, the Paris Disneyland is the most visited tourist destination not only in France but in all of Europe. I've often wondered why Disney selected France for the European operation (perhaps the company at least knew that a pantless Donald Duck would not be banned! see chapter 44). Disney parks succeed because they cater to kids and nostalgia and because they're some of the world's best-run service operations. Perhaps textbook authors will eventually come around, too.

Parking lots, called "car parks" in some areas, are the kind of park that many think of when they try to answer the question about where to find ninety thousand impalas. I didn't, however, have Chevies in mind but rather the large antelope found in Africa. Significant areas of Africa have been turned into national parks or preserves in the hope of saving wildlife from extermination. **Answer 45d:** Kruger National Park in South Africa is where you can find ninety thousand impalas. At least two of the large species in Kruger are threatened by poachers because they're worth huge amounts of money on a continent mired in poverty. Ivory from elephants is highly valuable, as is the "horn" of the rhinoceros, which is sold as an aphrodisiac.

Sadly, reports indicate that some of the leading poachers in Africa (not necessarily in Kruger) are people who are hired to protect the herds. Governments in Africa face the challenge of convincing their populations that wild animals are part of their national heritage and constitute a valuable resource to attract big-spending tourists.

CLIMATE AND
WEATHER

....

Question 46a: Where is the rainiest place on earth?

Question 46b: Where is the coldest place on earth?

Question 46c: Where is the hottest inhabited place on earth?

Question 46d: Where is the driest place on earth?

Question 46e: Where is the worst weather found?

....

The first geography class I ever took in college was a general course touching upon elements of both physical and human geography. The first topic was climate, which I fear represented the textbook author's belief that climate determined human behavior. The major focus of the chapters on climate was the Koppen system of climatic classification. We were expected to learn this system, an exercise that I found mind-numbing. I could see no rhyme nor reason to the system; it involved a bunch of capital and lowercase letters that I memorized for an exam, then forgot them as soon as possible. A few years later, I took a climatology course and realized I was again going to have to face the insane Koppen system. In his first lecture, the professor casually mentioned that Koppen was based on vegetation—in other words, the expression of climate on the earth's surface was really about "what grew where." As if giving a rationale for the system wasn't enough, we were then given actual data from climatic stations and had to work out the Koppen classification for each station. The result of this exercise was that it became impossible to forget the Koppen system.

The fact that a major climatic classification system is based on vegetation implies how we ought to consider the current matter of climatic change. Climate is a generalization about weather, but the nightly TV weather reports don't give us much of a perspective about long-term changes in climate. We know from historical records that crops (like traditional wine grapes) could be grown farther north in the early Middle Ages, then their area of cultivation retreated south during the so-called little ice age that lasted from around 1650 to 1850. Interestingly, the Maoris in New Zealand apparently faced a cooling trend there beginning around 1400. The Maoris were already challenged by New Zealand's climate, which was too cold to allow Polynesian staples like coconut and breadfruit to flourish. While modern plant science has developed varieties that will grow in a wide range of climatic conditions, native vegetation continues to give us a reliable indicator of climatic change.

Geographic trivia questions dealing with climate and weather most often deal with extremes. This is unfortunate because we measure weather in only a relatively few locations. Rainfall in particular can vary tremendously over short distances, so finding, for example, the rainiest place on earth is a very difficult undertaking. Despite this, it's actually the most common trivia question concerning weather and climate! Our search for the rainiest place involves looking for a location about four thousand to six thousand feet above sea level that regularly gets orographic rainfall. That is, the location faces the direction of the prevailing winds and its altitude forces the air flow to rise, thus cooling it and often causing it to rain. The location should also be one that can catch periodic storms that bring heavy rainfall. **Answer 46a:** One such location, and one often cited as the rainiest place on earth, is Cherrapunji, India, with annual rainfall of over 460 inches. A few places near Cherrapunji claim that their rainfall is slightly greater (which is certainly possible). Another contender for the title is Mount Waialeale on the island of Kaua'i in the Hawaiian Islands. The annual rainfall is slightly less than at the Indian locations, but in 1982 Waialeale received over 680 inches of rain! A meteorologist colleague of mine has told me that informal measurement on the windward slope of Haleakala Volcano, Maui, Hawai'i, reveals rainfall even greater than that on Waialeale.

Finding the coldest place on earth is equally elusive. One of the problems is that common thermometers aren't equipped to handle extremes. Twice in my lifetime I've been in temperatures of minus 40 degrees . . . I think. The (cheap) thermometers that measured these temperatures had a low possible reading of minus 40 and they were both pegged at their lowest readings. Incidentally, you may have wondered why I didn't report my personal "lows" in either Fahrenheit or Centigrade. It doesn't make any difference: at minus 40, both scales are the same! The coldest temperature ever recorded was, not surprisingly, in Antarctica, where a low of about minus 128 degrees Fahrenheit has been reported. We can certainly agree that's cold, but the problem is that the Antarctic is no one's home. What about inhabited areas of the world? **Answer 46b:** The town of Oymyakon in Siberia (then the Soviet Union) reached a low of about minus 96 degrees Fahrenheit.

Finding the hottest place is a bit easier. We can eliminate volcanoes, geothermal sites, and other places where temperature isn't related to weather. The hottest places are deserts. **Answer 46c:** The hottest reported temperature was in Libya, North Africa, where it reached approximately 136 degrees Fahrenheit. A more reliable reading was made in Death Valley, California, of about 134 degrees Fahrenheit in 1913. Of all the reported high temperatures in the world, the most surprising to me was the high of over 118 degrees Fahrenheit in Athens, Greece, in 1977. Athens isn't in a desert, and temperatures of that magnitude outside a desert region are truly unusual.

Answer 46d: There's no dispute about the driest place on earth. It is in Chile's Atacama Desert, where in some spots rain has never been recorded. This can be confirmed by noting that there are deposits of water-soluble compounds in up-slope locations in the Atacama but no trace of them down-slope. One of these minerals is sodium nitrate, commonly found in the Atacama. At one time it was a valuable resource for the country since it was used for, among other things, munitions. Around the time of World War I, the Germans discovered a process (the Haber process) by which ammonia could be manufactured at commercial volumes directly from the atmosphere. In turn, this allowed the manufacture of synthetic nitrates, which diminished the value of the Atacama's deposits.

Dry areas also produce a weather phenomenon known as dust storms, or sandstorms. Until recent concern with greenhouse gases, dust storms were considered to be the most significant generator of global climatic change, and their increase was blamed on poor management of dryland agriculture. In terms of their effect on topsoil and human livelihood, some of the greatest dust storms ever observed were a product of the dust bowl in the United States in the 1930s. These storms were created by mismanagement of lands in Oklahoma, Texas, and adjacent areas where grasslands in relatively dry areas were plowed for crops. In general, however, the worst dust storms are found in mid-latitude deserts like the Sahara.

The mechanics that generate a sandstorm help explain how the atmosphere can contain so much material from the ground that the sky can darken. Once particles are set in motion by the wind, they bounce (a process called *saltation*), which loosens larger particles. More recently it has been discovered that the dust particles contain an electric charge that's opposite to the charge of the ground itself. The particles are thus repulsed from the ground and essentially suspended in the atmosphere.

Here's a bit of bonus trivia: dry areas also produce a weather event known as "virga." This occurs when rain begins to fall but the air is so dry that the rain evaporates before it reaches the ground. Virga often creates severe instability in the atmosphere, usually in the form of downdrafts.

A number of news stories over the years have nominated places as having the worst weather. Places like Antarctica and Greenland are never mentioned, I guess, because the reporters haven't been there. The most recent place I've seen mentioned is Fargo, North Dakota, although Fargo was named primarily because of flooding, which really has as much to do with a drainage system as it does with weather. Previously I've seen Oklahoma City mentioned, probably because it's in the center of North America's Tornado Alley. **Answer 46e:** Among places nominated for bad weather, my personal favorite is Mount Washington, New Hampshire. Mount Washington is the highest peak in the northeastern United States, and the weather is so bad there that buildings are literally chained to the mountain. Wind gusts of over two hundred miles per hour have been recorded. Below-zero winter temperatures combined with high winds produce wind chill factors on Mount Washington that rival weather conditions in Antarctica and Greenland!

Some geographic research (and some introductory textbooks) have stressed the idea of "personal geography," how an individual learns about and responds to the environment. While the emphasis on the individual is interesting, I've often wondered how the authors of the articles or texts select the personal geographies they publish—is it actually the individual's geography I see in the articles or textbooks, or is it the authors' ideas of what they think I should be seeing?

While I haven't yet answered this question, it has occurred to me that I also have a personal geography when it comes to weather. As bad as it may be weatherwise in Fargo or on Mount Washington, I haven't spent enough time in either location to have a personal geography that includes their weather. On the basis of my activity space, I nominate Oswego County, New York, for the worst weather. Oswego County is on the southeastern shore of Lake Ontario and subject to what is now called "lake effect" snowfall. It's not so much the weather extremes there that warrant the nomination (although they can be noteworthy) as it is the surprises. On two occasions, I've parked my car in Oswego County in the early evening with no snow on the ground and returned at 11 p.m. to find snow accumulated to the handle on my car door. On one of these occasions, the snow continued to fall until it reached the roofline of some homes in Oswego, forcing the evacuation of residents.

NBC's *Today* show "weatherman," Al Roker, has helped Oswego County gain a national reputation. He attended college there at the State University of New York at Oswego and is personally acquainted with the weather. One Oswego County village he's mentioned on *Today*—Parish, New York—has a tendency to be buried in snow each winter. For all I know, its residents may hibernate until spring, a sort of weather-induced Brigadoon!

Snow isn't all that leads me to nominate Oswego County for the worst weather. One day I parked in a City of Oswego municipal lot. The local radio station reported that the temperature was 2 degrees below zero, Fahrenheit. While I was still sitting in the car, it began to rain—not a light rain but a heavy downpour. Showing rare presence of mind, I left the car and ran to shelter. Within five minutes, my car and every other one in the lot was encased in several inches of ice. Tree branches in the immediate vicinity began to break and fall from the heavy encrustation of ice. Utility lines had no chance. Oswego County is delightful in the summer . . . but that doesn't make up for its winter weather!

INDEX

ABOUT THE AUTHOR

Dr. Gary Fuller earned his PhD in geography at Penn State and taught for thirty-five years at the University of Hawai'i. He also taught introductory geography at Penn State and Ohio State. He was named a Teacher of the Year by the National Association for Geographic Education, captained a championship College Bowl team, and was a winning contestant on the television program *Jeopardy!* He served for thirteen years as a consultant to the US government on matters concerning population and national security. For the past twenty-five years he's lectured on geographic and population topics aboard cruise ships.